XXIII AIVELA Annual Meeting 2015

Journal of Physics: Conference Series Volume 778

Perugia, Italy
12 – 13 November 2015

Editor:

Enrico Primo Tomasini

ISBN: 978-1-5108-3532-0
ISSN: 1742-6588

Printed from e-media with permission by:

Curran Associates, Inc.
57 Morehouse Lane
Red Hook, NY 12571

Some format issues inherent in the e-media version may also appear in this print version.

Printed by Curran Associates, Inc. (2017)

For permission requests, please contact the Institute of Physics
at the address below.

Institute of Physics
Dirac House, Temple Back
Bristol BS1 6BE UK

Phone: 44 1 17 929 7481
Fax: 44 1 17 920 0979

techtracking@iop.org

Additional copies of this publication are available from:

Curran Associates, Inc.
57 Morehouse Lane
Red Hook, NY 12571 USA
Phone: 845-758-0400
Fax: 845-758-2633
Email: curran@proceedings.com
Web: www.proceedings.com

Table of contents

Volume 778

XXIII AIVELA (the Italian Association of Laser Velocimetry and non invasive diagnostics) Annual Meeting 2015

12–13 November 2015, Perugia, Italy

Accepted papers received: 24 November 2016Published online: 13 January 2017

Preface

Papers

IOP Conf. Series: Journal of Physics: Conf. Series **778** (2017) 011001 doi:10.1088/1742-6596/778/1/011001

PROCEEDINGS OF THE XXIII A.I.VE.LA. ANNUAL MEETING
Perugia, Italy, 12-13 November 2015
http://www.aivela.org/XXIII_Convegno/index.html

EDITORIAL

A.I.VE.LA. – the Italian Association of Laser Velocimetry and non-invasive diagnostics is a non-profit cultural association whose objective is to promote and support research in the field of non-contact or minimally invasive measurement techniques, particularly of electromagnetic-based techniques and optical techniques.

Through its Annual Meeting, AIVELA aims at creating an active and stimulating forum where current research results and technical advances can be exchanged and the development of new systems for laboratory use, field testing and industrial application can be promoted. The techniques covered include Laser Doppler Anemometry – LDA, Phase Doppler Anemometry – PDA, Image Velocimetry – PIV, Flow visualization techniques, Spectroscopic measurement techniques (LIF, Raman, etc.), Laser Doppler Vibrometry – LDV, Speckle Pattern Interferometry – ESPI, Holographic techniques, Shearography, Digital Image Correlation – DIC, Moiré techniques, Structured light techniques, Infrared imaging, Photoelasticity, Image based measurement techniques, Ultrasonic sensing, Acoustic and Aeroacoustic measurements, etc.

The first Annual Meeting was held back in October 1992 and since then the event has met large consensus among the research and scientific communities worldwide for the high scientific interest of the papers presented.

The XXIII AIVELA Annual Meeting was held at the premises of TUCEP- Tiber Umbria Comett Education Programme on 12-13 November 2015 and was organised in collaboration with the University of Perugia and TUCEP.
This volume contains a selection of the papers presented at the event.

The detailed Programme of the Meeting can be found at:
http://www.aivela.org/XXIII_Convegno/index.html

Trusting our Association and its initiatives will meet your interest, I wish to thank you in advance for your kind attention and hope to meet you soon at one of our events.

EDITOR
Enrico Primo Tomasini
President of A.I.VE.LA.

IOP Conf. Series: Journal of Physics: Conf. Series **778** (2017) 011002 doi:10.1088/1742-6596/778/1/011002

Peer review statement

All papers published in this volume of *Journal of Physics: Conference Series* have been peer reviewed through processes administered by the proceedings Editors. Reviews were conducted by expert referees to the professional and scienti c standards expected of a proceedings journal published by IOP Publishing.

IOP Conf. Series: Journal of Physics: Conf. Series **778** (2017) 012001 doi:10.1088/1742-6596/778/1/012001

Heart Rate assessment by means of a novel approach applied to signals of different nature

G Cosoli[1], **L Casacanditella, EP Tomasini, L Scalise**
Department of Industrial Engineering and Mathematical Sciences, Università Politecnica delle Marche, via Brecce Bianche, 60033 Ancona, Italy

E-mail: g.cosoli@univpm.it

Abstract. Electrocardiographic (ECG) signal presents many clinically relevant features (e.g. QT-interval, that is the duration of the ventricular depolarization). A novel processing technique has been demonstrated to be capable to measure some important characteristics according to the morphology of the waveform. Basing on that, the aim of this work is to propose an improved algorithm and to prove its efficacy in the assessment of the subject's Heart Rate (HR) in comparison to standard algorithms (i.e. Pan & Tompkins). Results obtained in experimentally collected ECG signals for the identification of the main feature (R-peak) are comparable to those obtained with the traditional approach (sensitivity of 99.55% and 99.95%, respectively). Moreover, the use of this algorithm has been broaden to signals coming from different biomedical sensors (based on optical, acoustical and mechanical principles), all related to blood flow, for the computation of HR. In particular, it has been employed to PCG (Phonocardiography), PPG (Photoplethysmography) and VCG (Vibrocardiography), where standard algorithms could not be widely applied. HR results from a measurement campaign on 8 healthy subjects have shown, with respect to ECG, deviations (calculated as 2σ) of ±3.3 bpm, ±2.3 bpm and ±1.5 bpm for PCG, PPG and VCG, respectively. In conclusion, it is possible to state that the adopted algorithm is able to measure HR accurately from different biosignals. Future work will involve the extraction of additional morphological features in order to characterise the waveforms more deeply and to better describe the subject's health status.

1. Introduction

Nowadays, the monitoring of physiological parameters is a fundamental issue not only in hospital environment, but also in other different application fields, from physical activity assessment (1,2) to e-health applications (3,4), such as in systems designed for the elderly care (5–7), home care (8–11) and wearable devices (12–14). Important quantities to assess are the Heart Rate (HR), the Respiration Rate (RR), the arterial blood pressure values and the blood oxygen saturation.

In particular, Electrocardiography (ECG) is the gold standard technique to evaluate the performance of the cardiovascular system (15,16); its typical waveform shows many clinically relevant features (e.g. the PR interval represents the duration of the atria depolarization (17)). According to the review in (18), HR can be measured by means of "contact", "fixed-in-environment" and "non-contact" sensors. In particular, among these devices the authors have considered the following ones:

[1] To whom any correspondence should be addressed.

- Phonocardiograph (PCG), which is a contact sensor, furnishing an acoustic signal acquired by means of a cardiac microphone placed on the subject's thorax, in proximity of the heart apex (19);
- Photoplethysmograph (PPG), a contact sensor too, providing an optical signal measuring the changes in light absorption by means of a photodiode placed in a standardized area illuminated by a LED; its signal is related to the blood flow and is similar to a typical arterial pressure waveform (20);
- Laser Doppler Vibrometer (LDV), providing a vibrocardiographic signal (VCG) consequent to the skin vibrations due to the blood flowing in the underlying vessel (21–23).

By considering the R-peak, ECG allows to easily assess HR, a fundamental parameter to monitor the subject's health status (15), which can be affected by several conditions (e.g. anxiety, stress or illness (16,24)). As regards the other signals (i.e. PCG, PPG and VCG), the fundamental aspect is to extract a periodic feature correlated with the heartbeat:

- first sound (S1) in PCG signal, corresponding to the instant when atrioventricular valves close at the beginning of systole (25);
- the principal peak (P-peak) in PPG, representing the variation in blood volume caused by the pressure pulse of the cardiac cycle (20);
- the main peak (V-peak) in VCG, caused by the same physiological principle as P-peak.

In 'figure 1' there are reported the typical biological signals acquired by the above mentioned sensing technologies and the localisation of their main features allowing the computation of HR.

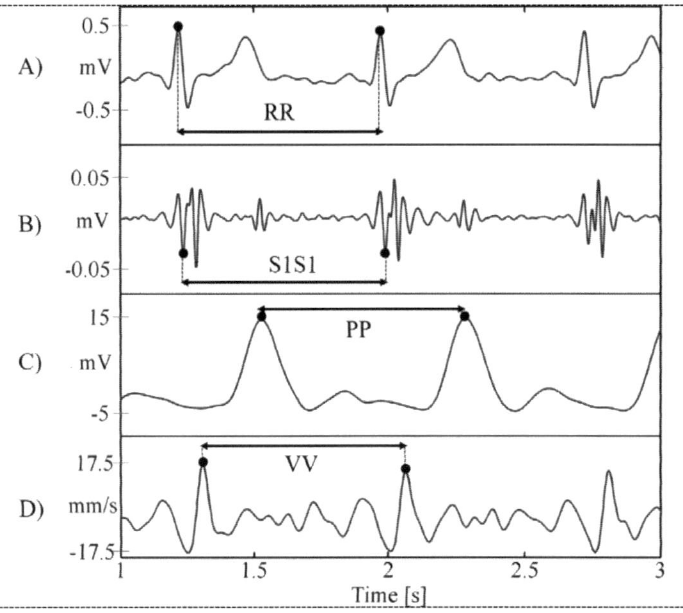

Figure 1. Typical biological signals and their main peaks to compute HR: A) ECG with RR interval; B) PCG signal with S1S1 interval; C) PPG signal with PP interval; VCG signal with VV interval

There are standard algorithms to detect ECG main feature, that is QRS complex (i.e. ventricular depolarization (26)). Pan&Tompkins (P&T) algorithm (27) has been the main processing procedure for ECG waveform for many years. However, it cannot be used neither to identify features different from QRS complex, nor to successfully characterize signals different from ECG (i.e. photoplethysmographic signal). Recently other algorithms have been developed, such as the one proposed by Hu et al. (28), based on the morphology of the signal waveform.

The aim of this work is to modify this algorithm and to adapt it in order to fully characterize signals of different nature, all related to blood flow (18): ECG, PCG, PPG and VCG signals. The authors want to evaluate its performance in terms of HR computation in comparison with the one of P&T algorithm, computing sensibility and positive predictivity indices (29).
In Materials and Methods paragraph the measurement setup and the processing technique are described; the results are reported in the following paragraph (Results) and discussed in the Discussion and Conclusion paragraph.

2. Materials and methods

2.1. Measurement setup

Eight healthy subjects (4 males, 4 females, aged 25±3 years, weight 61±13 kg and height 1.71±0.09 m) were investigated about their cardiac activity by means of the four sensing devices mentioned in the previous paragraph:

- ECG (ADInstruments MLA2540 5), II-lead;
- PCG (ADInstruments MLT201);
- PPG (Pulsed Amped Sensor);
- VCG (PDV 100, Polytec).

Such physiological signals were acquired by means of two proper 12-bit A/D acquisition boards: PowerLab 4/25 (ADInstruments) and NI 6008 (National Instruments). They were synchronized by means of the VCG signal, acquired to both the boards; in fact, with the first A/D board, necessary for the ECG acquisition, it is not possible to acquire all the sensors.
The sampling frequency was equal to 1 kHz and no digital filters were applied on the raw data; three 1-minute tests were made on each subject.
The measurement setup is reported in 'figure 2'.

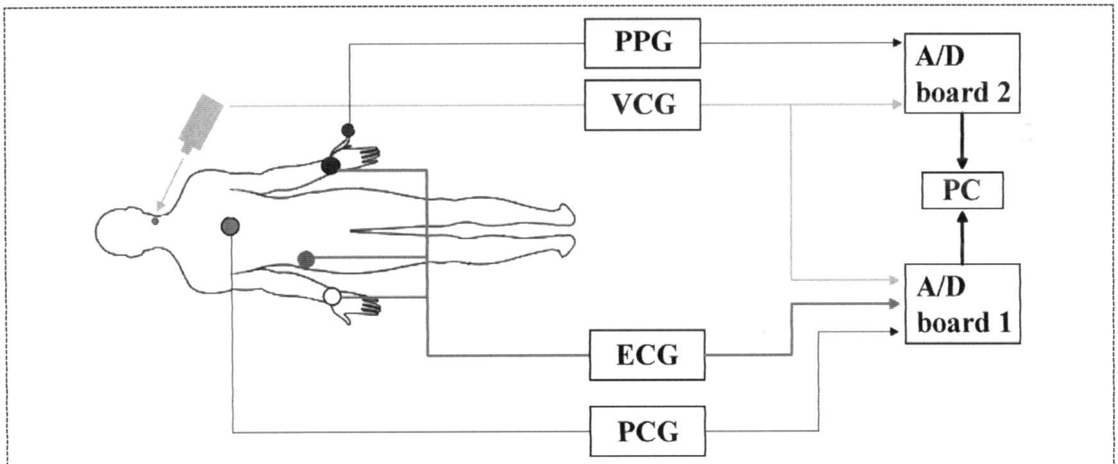

Figure 2. Measurement setup: ECG electrodes were placed on the wrists and on the hip bone (reference signal); PCG microphone was fixed on the subject's thorax near the heart apex; PPG sensor was applied on the left thumb; LDV was pointed in correspondence of the carotid artery (at a distance of 1 m)

The participant was asked to relax and stay supine during the acquisition. The ECG electrodes (disposable electrodes) were applied on the wrist and on the hip bone according to the standard Eindhoven triangle configuration (17). The cardiac microphone was fixed with a medical sticking plaster near the heart apex on the subject's thorax; PPG sensor was applied on the left thumb of the subject. Finally, the LDV was placed on a tripod, with the laser beam pointed perpendicularly on the subject's skin surface in correspondence of the carotid artery (at a distance of 1 m), which was spread with an hydrating lotion (45% zinc oxide) in order to maximize its reflectivity.

2.2. Data processing

The first step in data processing was the mean removal and filtering of the signals. Butterworth 3rd order band-pass filters were used, with the following cut-off frequencies: 0.4-20 Hz for ECG, 10-30 Hz for PCG, 0.4-30 Hz for PPG and 5-15 Hz for VCG signals. Such cut-off frequencies were chosen by observing the waveform before and after the filtering process: the goal was to highlight the main feature on each signals, removing the noise without lose useful information.
To identify the main feature on each signal, an algorithm based on (28) was developed. The approach consists in the evaluation of the slope trend in a portion of the signal defined by a mobile window, whose length is set equal to the 90% of heart period computed in the first 5 s of the ECG signal (by identifying the R-peaks manually). A 70% overlap of adjacent windows was applied in order to avoid missed beats.
With regard to ECG signal, HR was computed both by means of the standard P&T algorithm and with the proposed method. In the other signals, only the second approach was applied, since P&T is not effective.
The evaluation of the algorithm performance was done by considering two indices: Sensitivity (Se) (1) and Positive Predictivity (+P) (2).

$$Se = \frac{TP}{TP + FN} \tag{1}$$

$$+P = \frac{TP}{TP + FP} \tag{2}$$

where:

- TP are the True Positives, that is the number of correctly identified peaks
- FN are the False Negatives, that is the number of missed peaks
- FP are the False Positives, that is the number of wrongly identified peaks

The desirable situation is that Se and +P are both equal to 1, which means that all peaks are correctly identified and there are not false detections.

3. Results

HR results from this study on 8 healthy subjects are reported in this paragraph.
First, a comparison between Pan&Tompkins algorithm and the proposed approach is made in ECG signal. In 'table 1' there are FP, FN, TP, Se and +P mean values with the related standard deviation computed in the tested population.
The proposed approach was applied to all the acquired signals to compute HR. The results and the related percentage deviation of PCG, PPG and VCG with respect to ECG (i.e. gold standard) are reported in 'table 2' in terms of Heart Periods (HP), to better highlight the deviations with respect to ECG. Mean value and standard deviation computed in the three tests on each subjects are reported.

Table 1. Comparison between the performances of P&T algorithm and the proposed approach – ECG signal

Method	FP [m (σ) %]	FN [m (σ) %]	TP [m (σ) %]	Se [m (σ) %]	+P [m (σ) %]
P&T	0.34 (0.20)	0.06 (0.08)	99.94 (10.49)	99.95 (0.06)	99.64 (0.21)
Proposed approach	0.28 (0.31)	0.39 (0.30)	99.61 (10.63)	99.95 (0.36)	99.68 (0.38)

Table 2. HP [ms] computation by means of the novel approach in the acquired signals and percentage deviations (%d) of PCG, PPG and VCG with respect to ECG

	HP from the different sensing devices [ms]			
Subject	**ECG**	**PCG (%d)**	**PPG (%d)**	**VCG (%d)**
1	674	675 (0.1)	676 (0.1)	674 (-0.2)
2	725	722 (-0.4)	723 (0.1)	724 (0.0)
3	596	600 (0.7)	598 (0.4)	596 (0.4)
4	968	967 (-0.1)	966 (-0.1)	968 (0.3)
5	1028	1028 (0.1)	1028 (0.0)	1028 (0.0)
6	766	769 (0.3)	767 (-0.2)	765 (-0.2)
7	1112	1106 (-0.5)	1115 (0.8)	1112 (-0.2)
8	774	776 (0.2)	777 (0.2)	774 (-0.5)

Considering HR [bpm] (i.e. the inverse of HP [s], multiplied by 60, it is possible to state that the deviations (calculated as 2σ) are equal to ±3.3 bpm, ±2.3 bpm and ±1.5 bpm for PCG, PPG and VCG, respectively.

Moreover, the proposed algorithm allows to fully characterize a signal; an example of the identification of all the characteristic peaks of ECG waveform is reported in 'figure 3'.

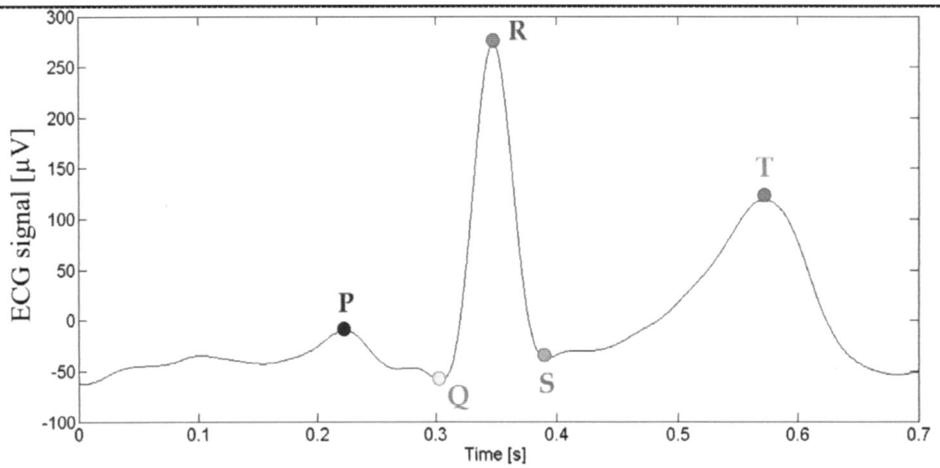

Figure 3. Example of the morphological characterisation of an ECG waveform by means of the proposed algorithm: R-peak, that is the main peak; P-wave, corresponding to the atria depolarization; Q-wave and S-wave, forming (together with R-peak) the QRS complex of the ventricles depolarisation; T-wave, representing the ventricles ripolarisation

4. Discussion and conclusions

The performance of the proposed approach in computing HR (or, equivalently, HP) from ECG signal is comparable to the one of the standard P&T algorithm. In fact, the adopted method reports a sensitivity of 99.95% and a positive predictivity of 99.68%, comparable to 99.65% and 99.64% reported by P&T, respectively.

An advantage of the proposed approach is the effectiveness in the morphological characterization of non-ECG signals, such as PCG, PPG and VCG (but, in the future, its application could be also widened to different ones). In fact, in HR computation deviations of ±3.3 bpm, ±2.3 bpm and ±1.5 bpm are reported for PCG, PPG and VCG, respectively, with respect to ECG (i.e. gold standard technique for HR measure).

In conclusion, it is possible to state that the proposed approach is effective in the measurement of HR from signals of different nature (i.e. ECG, that is an electric signal; PCG, that is an acoustic signal; PPG, that is an optical signal; VCG, that is a mechanical signal).

In addition, it allows to identify not only the main feature of a signal, but also other morphological characteristics ('figure3').

In the future, it would be interesting to extend the use of this algorithm, for example to identify different features in the acquired signals, looking for correlations in clinically relevant time intervals (e.g. to find an interval in VCG signal correlated to QT of ECG, to have an estimation of the duration of the ventricular depolarization obtained with a non-contact technique). In this way, the subject's health status could be described more in detail, providing an advantage above all if particular subjects have to be tested (e.g. non-contact measurements are particularly important in burnt patients or preterm infants, whose skin is particularly fragile and electrodes are difficulty placed (22)).

5. References

1. Bassett DR. Device-based monitoring in physical activity and public health research. Physiol Meas. 2012;33(11):1769.
2. Silva P, Santiago C, Reis LP, Sousa A, Mota J, Welk G. Assessing physical activity intensity by video analysis. Physiol Meas. 2015;36(5):1037.

3. G GD, da SRR, R BL, M O, M P. Monitoring heart rate variability online using e-health oriented 3G mobile telephone services. J Phys Conf Ser. 2013;477(1):12036.
4. Lymberis A. Wearable Ehealth Systems for Personalised Health Management: State of the Art and Future Challenges. IOS Press; 2004. 369 p.
5. Remote monitoring of health status of the elderly at home. A multidisciplinary project on aging at the University of New South Wales [Internet]. [cited 2016 Jul 18]. Available from: http://www.sciencedirect.com/science/article/pii/0020710195011396
6. Barger TS, Brown DE, Alwan M. Health-status monitoring through analysis of behavioral patterns. IEEE Trans Syst Man Cybern - Part Syst Hum. 2005 Jan;35(1):22–7.
7. Alexakis S, Karalis S, Asvestas P. Integrated system for remotely monitoring critical physiological parameters. J Phys Conf Ser. 2015;637(1):12002.
8. Wong MYM, Pickwell-MacPherson E, Zhang YT. Contactless and continuous monitoring of heart rate based on photoplethysmography on a mattress. Physiol Meas. 2010;31(7):1065.
9. Fensli R, Gunnarson E, Gundersen T. A wearable ECG-recording system for continuous arrhythmia monitoring in a wireless tele-home-care situation. In: 18th IEEE Symposium on Computer-Based Medical Systems (CBMS'05). 2005. p. 407–12.
10. Smart Monitoring of User's Health at Home: Performance Evaluation and Signal Processing of a Wearable Sensor for the Measurement of Heart Rate and Breathing Rate [Internet]. [cited 2016 Jul 18]. Available from: https://www.researchgate.net/publication/295859566_Smart_Monitoring_of_User's_Health_at_Home_Performance_Evaluation_and_Signal_Processing_of_a_Wearable_Sensor_for_the_Measurement_of_Heart_Rate_and_Breathing_Rate
11. Pietroni F, Casaccia S, Revel GM, Scalise L. Methodologies for continuous activity classification of user through wearable devices: Feasibility and preliminary investigation. In: 2016 IEEE Sensors Applications Symposium (SAS). 2016. p. 1–6.
12. Korhonen I, Parkka J, Gils MV. Health monitoring in the home of the future. IEEE Eng Med Biol Mag. 2003 May;22(3):66–73.
13. Zheng YL, Ding XR, Poon CCY, Lo BPL, Zhang H, Zhou XL, et al. Unobtrusive Sensing and Wearable Devices for Health Informatics. IEEE Trans Biomed Eng. 2014 May;61(5):1538–54.
14. Pantelopoulos A, Bourbakis NG. A Survey on Wearable Sensor-Based Systems for Health Monitoring and Prognosis. IEEE Trans Syst Man Cybern Part C Appl Rev. 2010 Jan;40(1):1–12.
15. Orso F, Baldasseroni S, Maggioni AP. Heart Rate in Coronary Syndromes and Heart Failure. Prog Cardiovasc Dis. 2009 Jul 1;52(1):38–45.
16. Custodis F, Schirmer SH, Baumhäkel M, Heusch G, Böhm M, Laufs U. Vascular pathophysiology in response to increased heart rate. J Am Coll Cardiol. 2010 Dec 7;56(24):1973–83.
17. Rautaharju PM, Surawicz B, Gettes LS. AHA/ACCF/HRS Recommendations for the Standardization and Interpretation of the Electrocardiogram. Circulation. 2009 Mar 17;119(10):e241–50.
18. Kranjec J, Beguš S, Geršak G, Drnovšek J. Non-contact heart rate and heart rate variability measurements: A review. Biomed Signal Process Control. 2014 Sep;13:102–12.
19. Walker HK, Hall WD, Hurst JW, editors. Clinical Methods: The History, Physical, and Laboratory Examinations [Internet]. 3rd ed. Boston: Butterworths; 1990 [cited 2015 Oct 28]. Available from: http://www.ncbi.nlm.nih.gov/books/NBK201/
20. Alian AA, Shelley KH. Photoplethysmography. Best Pract Res Clin Anaesthesiol. 2014 Dec 1;28(4):395–406.
21. Pinotti M, Paone N, Santos FA, Tomasini EP. Carotid artery pulse wave measured by a laser vibrometer. In: Third International Conference on Vibration Measurements by Laser Techniques: Advances and Applications [Internet]. International Society for Optics and Photonics; 1998 [cited 2015 Oct 12]. p. 611–616. Available from: http://proceedings.spiedigitallibrary.org/proceeding.aspx?articleid=954179

22. Cosoli G, Casacanditella L, Tomasini EP, Scalise L. Evaluation of Heart Rate Variability by means of Laser Doppler Vibrometry measurements. J Phys Conf Ser. 2015;658(1):12002.
23. Morbiducci U, Scalise L, De Melis M, Grigioni M. Optical vibrocardiography: a novel tool for the optical monitoring of cardiac activity. Ann Biomed Eng. 2007 Jan;35(1):45–58.
24. Fox K, Borer JS, Camm AJ, Danchin N, Ferrari R, Lopez Sendon JL, et al. Resting heart rate in cardiovascular disease. J Am Coll Cardiol. 2007 Aug 28;50(9):823–30.
25. Phonocardiography Signal Processing - Abbas K. Abbas, Rasha Bassam - Google Libri [Internet]. [cited 2016 Jul 18]. Available from: https://books.google.it/books?id=x4P25KZCAB8C&printsec=frontcover&dq=phonocardiography&hl=it&sa=X&redir_esc=y#v=onepage&q=phonocardiography&f=false
26. Catalano JT. Guide to ECG Analysis. Lippincott Williams & Wilkins; 2002. 518 p.
27. Pan J, Tompkins WJ. A Real-Time QRS Detection Algorithm. IEEE Trans Biomed Eng. 1985 Mar;BME-32(3):230–6.
28. Hu X, Liu J, Wang J, Xiao Z, Yao J. Automatic detection of onset and offset of QRS complexes independent of isoelectric segments. Measurement. 2014 May;51:53–62.
29. Arzeno NM, Deng ZD, Poon CS. Analysis of First-Derivative Based QRS Detection Algorithms. IEEE Trans Biomed Eng. 2008 Feb;55(2):478–84.

IOP Conf. Series: Journal of Physics: Conf. Series **778** (2017) 012002 doi:10.1088/1742-6596/778/1/012002

Eye tracker uncertainty analysis and modelling in real time

A. Fornaser[1,3], M. De Cecco[1,3], M. Leuci[1], N. Conci[2,4], M. Daldoss[4], A. Armanini[2,4], L. Maule[1], F. De Natale[2,4], M. Da Lio[1]

[1]Department of Industrial Engineering, University of Trento, via Sommarive, 9 Trento (TN) Italy.
[2]Department of Information Engineering and Computer Science, University of Trento, via Sommarive, 9 Trento (TN) Italy.
[3]Robosense - Salita dei Molini, 2 38123 Villazzano (TN)
[4]Xtensa srl - Salita dei Molini, 2 38123 Villazzano (TN)

mariolino.dececco@unitn.it

Abstract. Techniques for tracking the eyes took place since several decades for different applications that range from military, to education, entertainment and clinics. The existing systems are in general of two categories: precise but intrusive or comfortable but less accurate. The idea of this work is to calibrate an eye tracker of the second category. In particular we have estimated the uncertainty both in nominal and in case of variable operating conditions. We took into consideration different influencing factors such as: head movement and rotation, eyes detected, target position on the screen, illumination and objects in front of the eyes. Results proved that the 2D uncertainty can be modelled as a circular confidence interval as far as there is no stable principal directions in both the systematic and the repeatability effects. This confidence region was also modelled as a function of the current working conditions. In this way we can obtain a value of the uncertainty that is a function of the operating condition estimated in real time opening the field to new applications that reconfigure the human machine interface as a function of the operating conditions. Examples can range from option buttons reshape, local zoom dynamically adjusted, speed optimization to regulate interface responsiveness, the possibility to take into account the uncertainty associated to a particular interaction. Furthermore, in the analysis of visual scanning patterns, the resulting Point of Regard maps would be associated with proper confidence levels thus allowing to draw accurate conclusions. We conducted an experimental campaign to estimate and validate the overall modelling procedure obtaining valid results in 86% of the cases.

1. State of the art

The eye tracker is an instrument that allows to determine the point where the user is looking at. This kind of technology has two main fields of application: analysis of visual scanning patterns and human machine interfaces (HMI).

Since it is commonly agreed that the point or region a person is looking at is the point where that person is focusing his attention, except in some cases with very simple task as stated by Rayner and Keith in [1], knowing the visual scanning pattern it's a great way to study human cognitive processes. Applications may be several starting from market to web searches. Moreover the study of visual scanning path may be used in medicine to diagnose mood, perception, learning and attention disorders. Also in the field of human machine interface the use of the eye as input have multiple advantages. The first and more evident reason to use the eye as input, is to allow users that are not able to use other input, such as joystick and keyboard, due to some diseases that partially or completely limit their movement. Anyway the eye tracker may be useful to any kind of user. Indeed eye movement is one of the fastest human movement, although are mainly conceived for exploration, less for control. A good insight can be found in [2].

There are different technologies implemented in order to track the eye [3]. Through all this technologies it is possible at first to distinguish between intrusive and non intrusive one. In the first category falls all that kind of devices that are in some way fixed with the user's head. Some examples are: contact lenses, electrodes and head mounted devices. This kind of technology have two main advantages: accuracy and robustness with respect to head movement. The main drawback of intrusive devices is that may be tedious for the user. The alternative are the non intrusive, or remote, eye tracker [4]. This kind of devices are mostly based on computer vision techniques. They consist of one or more cameras to capture the eye's image. Those kind of system are more comfortable and can be used for long periods. This make them suitable to be used as HMI. Due to the scope of this work we are more interested in remote eye trackers.

The most popular kind of remote eye trackers use features detection to track the eye. Eye's characteristic that can be tracked are, e.g., the limbus and the pupil. The problem of the limbus is that it's covered by the eyelids. On the other hand the problem of the pupil is the weaker contrast with the iris, that makes it harder to detect it in the image. This will obviously depend also on the specific user characteristics, such as shape and size of the eye and also the color. Anyway a solution commonly adopted is the usage of infrared light. Indeed IR light enhance the above mentioned contrast. Since IR is not visible it won't distract the user.

The need of a common way to estimate eye trackers' performances has been discussed by Holmqvist and Nyström in [5]. In particular they determine three principle characteristic in order to determine the performances of the eye tracking devices:

- Accuracy: is the bias (**systematic effect**), i.e. the difference between the estimated Point of Regard PoR (mean of the collected data) and the point the user is actually looking at.
- Precision: that is the **repeatability** of the estimate. It is usually estimated as the RMS or as the standard deviation of the available data.
- Robustness: capability to work with different users. Usually eye trackers do not work with everybody or anyway do not have same performances changing users. This may be due to different characteristics of the user, both cognitive (e.g. the ability to follow a task) and physics (e.g. the physiology of the eye).

A common problem underlined from many authors is that the uncertainty provided by the manufacturers seems not to correspond to real cases. This is mainly due to the fact that the values obtained by the manufacturers are obtained in ideal conditions. Knowing the real accuracy and precision allow better design of HMIs [9].

In order to calibrate a commercial eye tracker we identified the following parameters of influence:

- Illumination;
- Relative motion between the user head and the instrument (rotation and translation);
- Objects in front of the eyes;
- Different user.

Many authors have tried to identify the influencing parameters in order to find a way to decrease their effect [9, 10, 11, 12, 13, 14]. It is commonly considered that two of the main problem of the remote eye tracking systems are illumination change and head movement.

Illumination: the influence of light largely depend on the usage of the eye tracker. For example if it is used as HMI on a computer it is possible to place that computer in a spot with optimal illumination. For example far from windows (illumination change during the day). Hansen and Pece in [9] propose a system that use image statistic instead of features detection. As described by the authors "the underlying idea is that a large image gradient is likely to arise from a boundary between object and background". Indeed they obtain a system that is robust with respect to light changes. They use an algorithm based on particle filters and EM Contour, the main drawback is that head movement have to be very limited.

Relative motion between the user and the instrument (rotation and translation): the problem of head movements will largely depend on the user characteristics: several disabled user may not be able to move the head or may not be able to keep it fixed depending on the diseases. Also a healthy user will slightly move the head in most activities. Usually to test the eye tracker it is used a chin rest or a bite bar that eliminate those kind of movement. The results obtained will be largely different from the ones achievable in real operating conditions. In literature multiple solutions to the head movement problem can be found: in [10] White, Hutchinson and Carley propose to include a second reference light source to distinguish between head and eye rotation. They proof the validity of the solution for large lateral displacement, but only with simulated data. Cerrolaza et al. in [11] instead consider the problem of head displacements in the direction perpendicular to the screen. In particular they proposed two methods. The first one is to include information about the different positions in the calibration procedure in order to improve the tolerance to this kind of displacement. A noticeable drawback for this method is the increase of the calibration time. For this kind of system a too long calibration time may increase the stress of the user and the distraction. The second method is based on the hypothesis that the PoR estimation error mainly depend on the system and not on the different user. Hence the error can be modeled for a particular configuration and used to compensate the estimation. Both method seems to improve the robustness to head movement. In [12] Guestrin and Eizenmanit stated that with more than one light source, as in the case of the system we are going to use, it is possible to tolerate certain movement of the head. That are from 2 to 4 cm depending on the direction of the displacement. That requires a multiple point calibration. In addition they show that the same results can be obtained with a single point calibration using both multiple light sources and multiple cameras. The idea of using more than one camera is exploited in more than one paper. For example in [14] multiple cameras and multiple lights are used to eliminate the usage of the user dependent parameters. The estimate is done using 3D computer vision technique and allows to completely eliminate the calibration phase. On the other hand in [13] the new camera is used in order to allow free head movement in a volume with 40 cm of diameter. This may result very interesting in the case of a user in front of a computer screen, since the natural movement won't be much bigger than that.

Anyway what we are actually interested in is not to change how the eye tracker works but to determine:

1. the accuracy estimated in the LCD plane in relation to the influencing factors;
2. the effect of the influencing factor on eye tracker accuracy and its associated covariance in real time.

The first goal is something similar to what Nystrom and al. reported in [6]. Here the authors consider calibration method and eye physiology as influencing parameters and study their effect on data quality. The test are done with a tower mounted video based eye tracker. In particular they study three difference calibration procedures to verify which one works better. The difference through the three is the way to determine if the calibration has to be considered good or discard. The three kinds are: operator controlled, participant controlled and automatic calibration. From the experimental data it seems that participant controlled calibration is the one that gives the better results, followed by operator controlled one. During the test both the user and the operator can see if the system is capturing the eye position in the right way. In addition they develop a series of test to verify how accuracy, precision and the number of valid fixation samples vary changing the following influences parameters: position on the screen, time, eye physiology and visual aids. In the following the results they obtain:

- Position on the screen: The off center positions correspond to an higher visual angle that makes it harder to distinguish the eye, e.g. the pupil may be covered by eyelashes.
- Time: Data quality get worst with time. This may be due to the fact that person move from his initial position, even if with the chin rest this should not be a problem. The problem of moving from calibration position will be discussed later also in this work.
- Eye physiology: Eyelashes pointing downwards and smaller pupil has a negative effect on accuracy. Precision is worst for blue eyes. This is mainly due to the fact that blue eyes have a lower contrast between iris and pupil.
- Visual aids: Worst result both with glasses and contact lenses. The problem with contact lenses is that they can slight with respect to the eye, while the issue with glasses is that they absorb some of the infrared light (depending on the material they are made of).

In this article the authors provide the first comprehensive set of data showing how the calibration method, the operator, participants' eye physiology, and visual aids affect the quality of data.

Based on the work of Holmqvist et al. [5], it has been developed a software 'Accuracy Test Tool', for precision and accuracy measure by Tobii Technology. In particular a description of the software and test specification can be found in [7]. As said by the authors the document presents a suitable methodology to test and compare the performance of different remote eye tracking systems. It outlines a series of extensive tests that identify and control for external parameters that illustrate the accuracy and precision of the system under different scenarios (e.g. subjects position in the track box, environmental light levels, and large gaze angles). Unfortunately in this paper they do not take into account the influence parameters that depend on user characteristics (eye physiology, glasses...). Indeed they consider only data acquired by "suitable eye tracking individuals". This means that people with glasses, lenses or anyhow poor performances are excluded. The different scenarios they consider are:

- Ideal conditions: best scenario, unattainable in real working conditions.
- Large gaze angles: stimuli placed far from center.
- Varying illumination: variation of illumination level.
- Head positions: the eye tracker is moved of 5 cm in one direction (X, Y or Z) for test. The head always remain on the chin rest.

It has to be noticed that in all this test the user's head lay on a chin rest. This is not what happened in real situation where the user may move the head. Another calibration procedure, based on this idea, was carried out by Clemotte et al. in [8]. The purpose of the work is to "identify the precision and accuracy of the Tobii x2-30 with non disabled people under non ideal conditions (without any chin rest)". They found out that accuracy(systematic effect) and precision (repeatability) values obtained in these conditions highly differ from the ones reported on the data sheets: the accuracy range from 0.4 to 2.46 degrees, while the precision range from 0.2 to 1.91 degrees.

Despite several works can be found regarding the device calibration, at the current state of the art there are no works proposing the estimation of the accuracy as a function of the actual conditions. We believe that this will open the scene to a more effective use of the device. In the analysis of visual scanning patterns the resulting PoR maps would be associated with proper confidence levels that for sure would represent a fundamental instrument to draw accurate conclusions. In the field of HMI examples can range from option buttons reshape, local zoom dynamically adjusted, speed to regulate interface responsiveness, the possibility to take into account the uncertainty associated to a particular interaction. This part represents the most innovative contribution of our work.

2. Experimental set up and test description

In the experimental campaign we adopted the EyeAssist developed by Xtensa, a low-cost eye tracking platform developed for clinical use3. Using an infrared camera, it maps the gaze of the user on the screen by detecting the location of the pupil with respect to a known pattern of light. This system requires a first phase of calibration to fit the different users. This phase consists in a point moving on the screen in a random way between 9 possible positions. In particular we target the PoR together with its covariance that we estimate as a function of the operating conditions in real time, namely the lighting, head position and rotation, target position on the screen, number of eyes detected and user characteristics.

Figure 1: Eye tracker system mounted on the screen

The eye tracker communicate data obtained on a UDP port. The software developed in Unity was able to read from this port and also to communicate messages to the eye tracker, always through the UDP. This to communicate the data needed to load the calibration. The script that allows the communication was obtained starting from [15].

During the test, the software counts both the valid and the wrong data. If the number of wrong data overcomes 130, the target is moved to the following position. The data obtained contain:

- Timestamp;
- X coordinate of the PoR obtained considering only the left eye;
- Y coordinate of the PoR obtained considering only the left eye;
- X coordinate of the PoR obtained considering only the right eye;
- Y coordinate of the PoR obtained considering only the right eye;
- X coordinate of the PoR obtained as a medium value;
- Y coordinate of the PoR obtained as a medium value;
- X position of the left pupil on the image;
- Y position of the left pupil on the image;
- X position of the right pupil on the image;
- Y position of the right pupil on the image;
- X coordinate of the target;
- Y coordinate of the target;
- Number of wrong data.

The test procedure used to collect the data is similar to the one used for calibration. The software employed to develop the test was Unity™. Unity is an integrated multiplatform to develop animations in many different fields. In our case we used it to control a circle on the screen to appear randomly on a grid of 5x5 that covers the whole screen. The circular target remained in each location on the screen until a sufficient number of valid data has been collected. This number was set to 130.

A problem that we considered was the relative motion between the screen and the head. In order to minimize this effect we added a bar to the set up to hold the user forehead. Moreover it has been drawn a scheme that allow the user to verify if he lies in the center of the camera in order to locate the head always in the same position. Both solutions can be seen in the following images, Figure 2 and Figure 3.

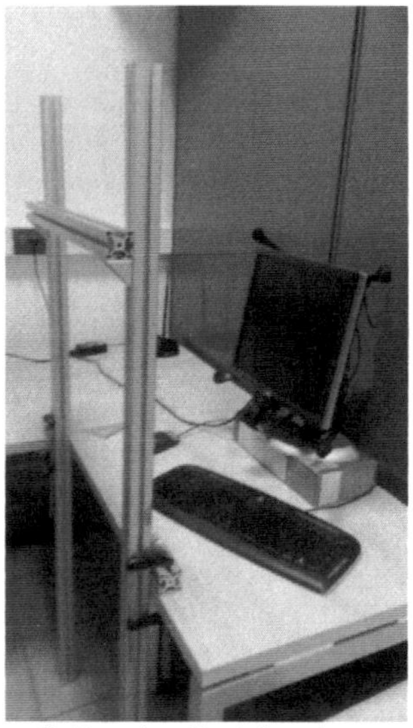

Figure 2: Bar that limit head movement

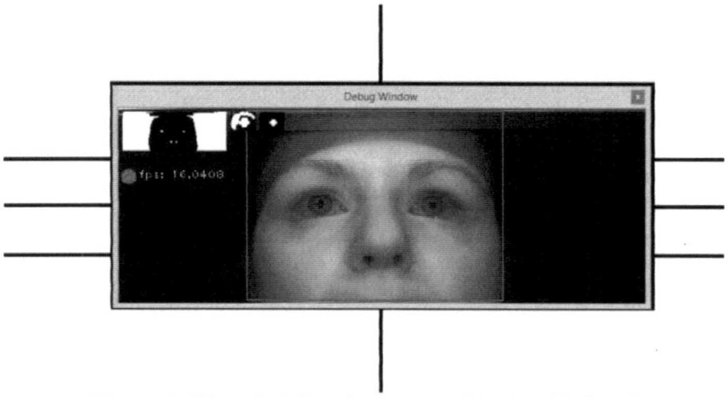

Figure 3: Lines helping the user positioning his head

Another influencing factor is the light. Indeed ambient light may change during the day. This alteration are partially compensated by changing the camera exposure. Anyway the test were conducted in as much limited time slots as possible, also verifying that the lighting condition didn't changed that much.

3. Eye tracker repeatability and accuracy characterization

The obtained data were elaborated to remove outliers by means of the Chauvenet's. We set the maximum allowable deviation to 2. The procedure was repeated until the difference between previous and current standard deviation was less than 10%. Higher reduction would be meaningless. In Figure 4 it is evident that the noise was largely reduced.

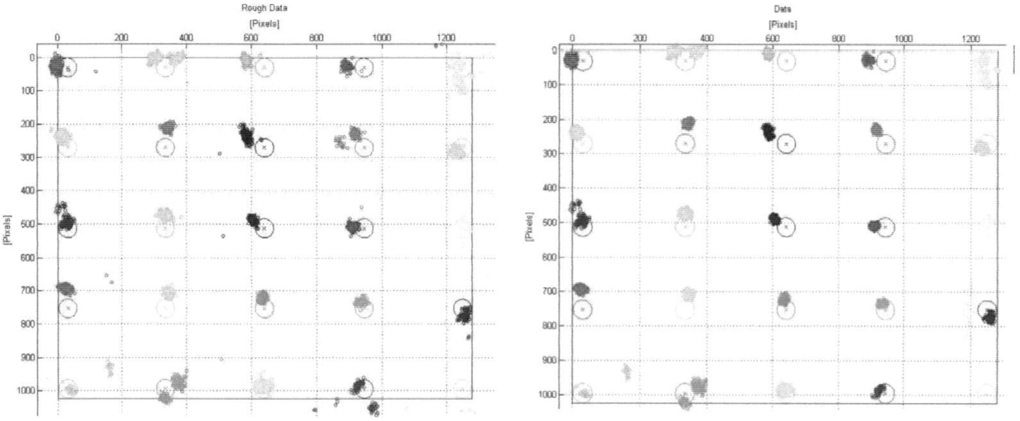

Figure 4: Data collected for accuracy and repeatability characterization, before (left) and after (right) the application of a filter

Having the filtered data we can calculate the systematic effect and repeatability for every target position. The repeatability regions were then calculated using a confidence threshold that allows to obtain a confidence level of 95%, as explained in [16]. Plotting the results obtained it is straightforward to see that the ellipse does not contain the true value, which is the target position.

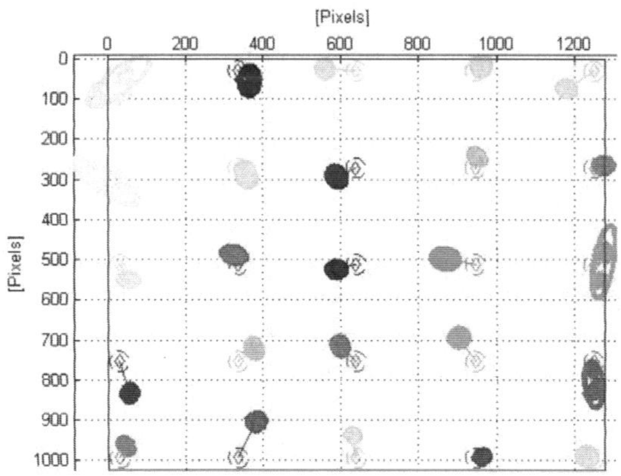

Figure 5: Mean and Covariance

It is evident that this problem is mainly due to the presence of a strong systematic effect that seems to have a random nature, as evident in figure 5. In traditional measurement systems, if the results are repeatable, the systematic effect can often be determined and compensated.

In order to verify the statistic nature of the systematic effect, we tried to distinguish between repeatability for the same calibration and repeatability through different calibrations. For this aim we decided to consider data coming from different calibrations. In figure 6 it is evident that the systematic effect is completely random.

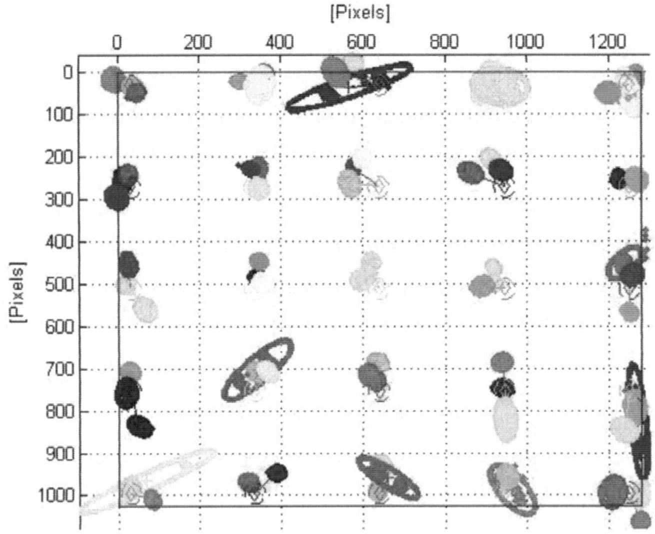

Figure 6: Repeatability through different calibrations

4. Eye tracker accuracy characterization

The results presented previously underline the need of a different method to estimate uncertainty. It has to be noticed that we have a good estimation of the true value, that is the target position. Even though we can't say where a person is actually looking but only where he thinks to look, the target position is a good approximation since the user is asked to look at the point. Hence we can use this information to calculate an interval that includes the true values. That is we can calculate the residuals as the difference between nominal value and measured values, and use those residuals to build the covariance matrix.

We use more than one test in order to have more data and also to have data coming from different calibrations. Indeed we have seen that data coming from different calibrations are not repeatable. Anyway all the data are still coming from the same user. Note that the covariance calculated with the distance from the target do not need a confidence threshold of two to achieve the 95% of confidence, as in the pure repeatability case, but a confidence threshold of one is more then enough because the repetitions are always concentrated several pixels away from the reference value.

The covariance (equiprobable) ellipse can be expressed by three parameters that represent the major, the minor axis and the orientation (i.e. correlation factor). Looking at the available data, see Figure 7, it is evident that the orientations are random.

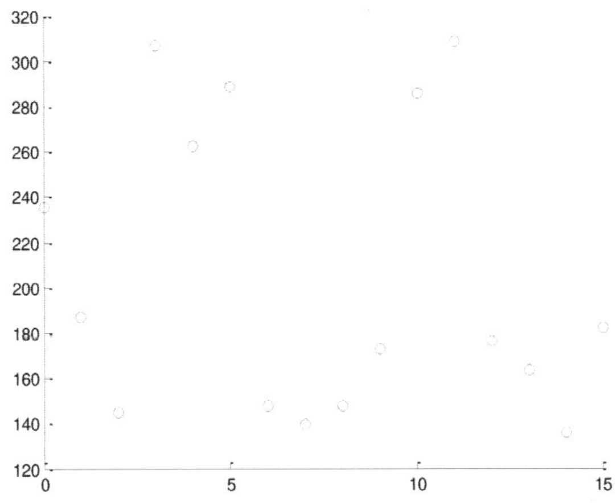

Figure 7: Values of the covariance ellipse's angle

Because there isn't a prevalent orientation of the equiprobable ellipses it has no sense to model it as an ellipse. As a consequence we modelled the uncertainty to be equally distributed along every direction, i.e. as a circle of radius R equal to the semi major axis. We found the value of R that contains the 95% of the values to be R=114 pixels. In Figure 8 can be seen this value applied to a validation set.

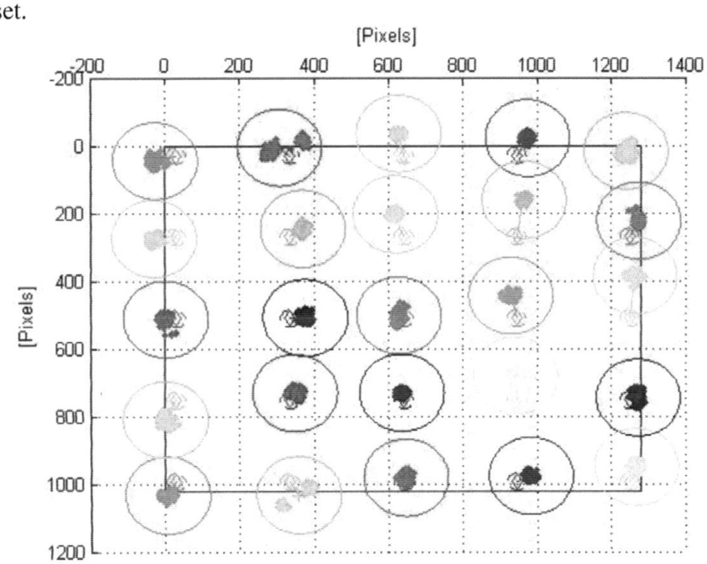

Figure 8: Covariance 'circle' with radius R=114 pixels

All the targets are now inside the covariance ellipse, that is actually a circle, except for one. Using a wider series of data we verify that in the 95% of the cases the circle include the true value. Since the screen used is 1280x1024, this value is more or less the 10% of the screen dimension.

5. Modelling of the uncertainty as a function of the Influencing Parameters in Real Time

In this chapter we are going to model the uncertainty circle as a function of the operating conditions. The influencing parameters that have an effect on the variance are the following:

- **Target position on the screen**: as already underlined in the previous section in different positions of the screen we have different behaviors of the eye tracker. For example in the extreme points of the screen is more probable to have the eye covered by eyelashes or eyelid.
- **Illumination**: in a system that is not moving and placed in an appropriate spot, there should be limited illumination changes. That is e.g. far from window. Anyway considering a mobile application there can be wide changes in illumination.
- **Eyes detected**: the eye tracker find both the PoR of right and of left eye. The estimate is a mean weighted considering which one is the dominant eye. In some cases, usually the most external part of the screen, the eye tracker may not be able to identify both. If the detected eye is not the dominant one performances may largely change.
- **Relative movement between user and instrument**:
  - Displacement: both in the direction parallel to the screen and in the direction perpendicular to the screen.
  - Rotation: sometimes to look somewhere a person move both the eyes and the head or maybe just the head. Rotation may lead to difficulties in detecting the eyes.
- **Object in front of the eyes**: like glasses, contact lenses and hair.
- **Different user**: in this category may fall a lot of characteristic, such as eye's color, the cognitive ability, shape of the eyes and so hence so forth. Note that in this work all the tests are done by the same person, this means that those factors won't be considered. It is clear that this may be an interesting topic for future research.

Through all this influencing parameters we decide to model only: target position on the screen, number of eyes and head movement with respect to the instrument.
In the test about the influencing parameters the statistical control for each of them is obtained varying only one at once. That is, e.g., changing only the illumination keeping all the other parameters as fixed as possible. The test used is the same explained previously. Those are the values that we can quantify in real time given the information we have access to. The effect of every parameter will be modelled independently thus neglecting correlations between them.

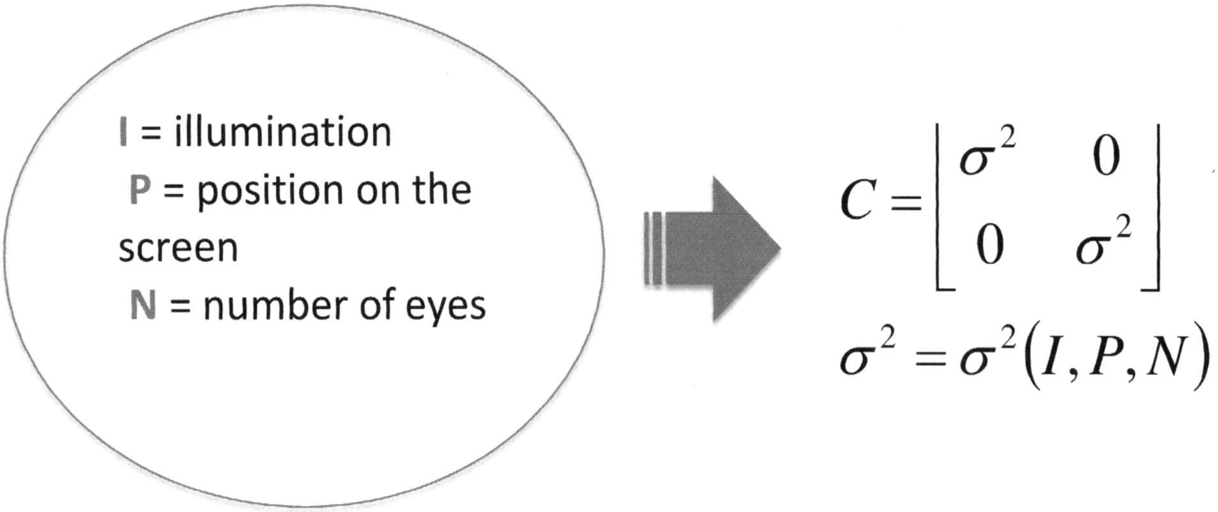

Figure 9: Influencing factors correlation/modelling wrt uncertainty

IOP Conf. Series: Journal of Physics: Conf. Series **778** (2017) 012002 doi:10.1088/1742-6596/778/1/012002

Note that the dependency from each parameter considered does not have to be linear. In the following we will find the relation between the parameters considered and the covariance, namely its radius R.

5.1. Target position on the screen.

The first influencing parameter we are going to consider is the target position on the screen. From the ANOVA test it comes out that it makes sense to divide the values along the diagonal, as can be seen from the following box plots, we will use the distance from the diagonal as a representation of the position.

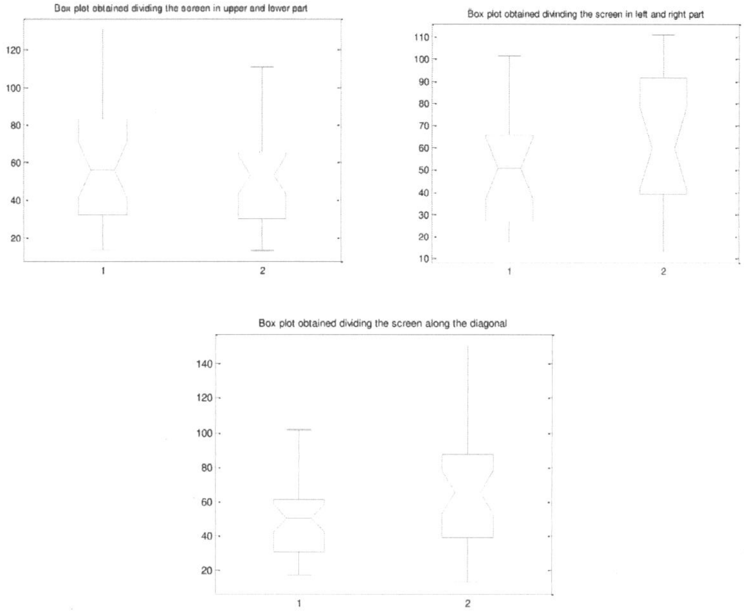

Figure 10: ANOVA test results

That is the influencing factor will be how far the point is from the diagonal, that is the line going from the upper left to the lower right part of the screen. This is probably due to the asymmetries of the system.

Taking the already considered three tests we model the value of the covariance radius R for every position and the corresponding distance from the diagonal. The linear model was computed as the line that covers the 95% of the radius. Hence the function will be:

$$\frac{dR}{dP} = 0.009$$

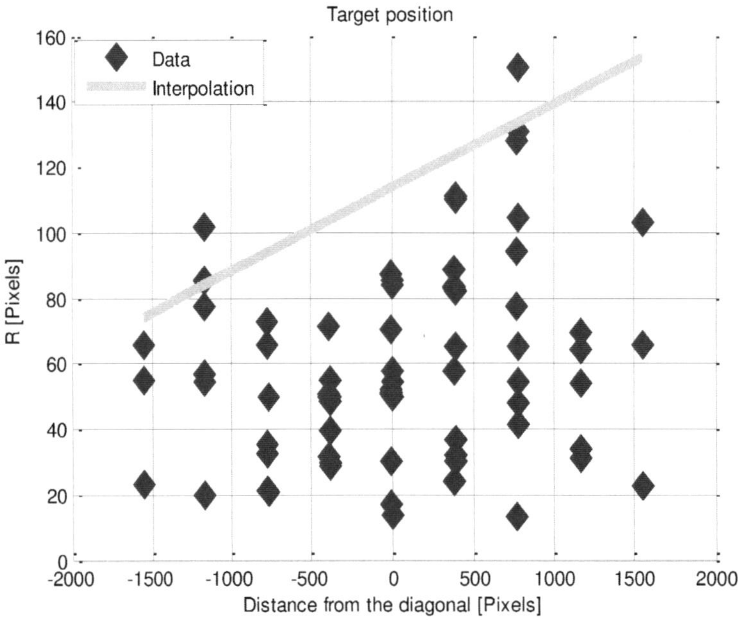

Figure 11: Target position on the screen.

5.2. Eyes detected

Unlike all the other parameters this is not a continuous value. We took three tests as before and consider the data coming from the right eye, the ones coming from the left eye and the ones coming from the mean. In Figure 12 we can see that the behaviour obtained with the information coming from only left eye strongly differs. In this case the user's dominant eye is the right one. Indeed R gets bigger when considering the data coming from only the left one. Hence we can find two values of R, one to be applied in the case we have the information from the right eye, or more in general the dominant eye, and one for the other cases. To calculate those values we consider a number that include the 95 % of the available data. The two values obtained, as can be seen in the figure, are 114 pixels for the right eye (blue line) and 188 pixels for the left eye (red line). So the model we finally obtain will be:

$$\frac{dR}{dN} = \begin{cases} 74 \ if \ left \\ 0 \ otherwise \end{cases}$$

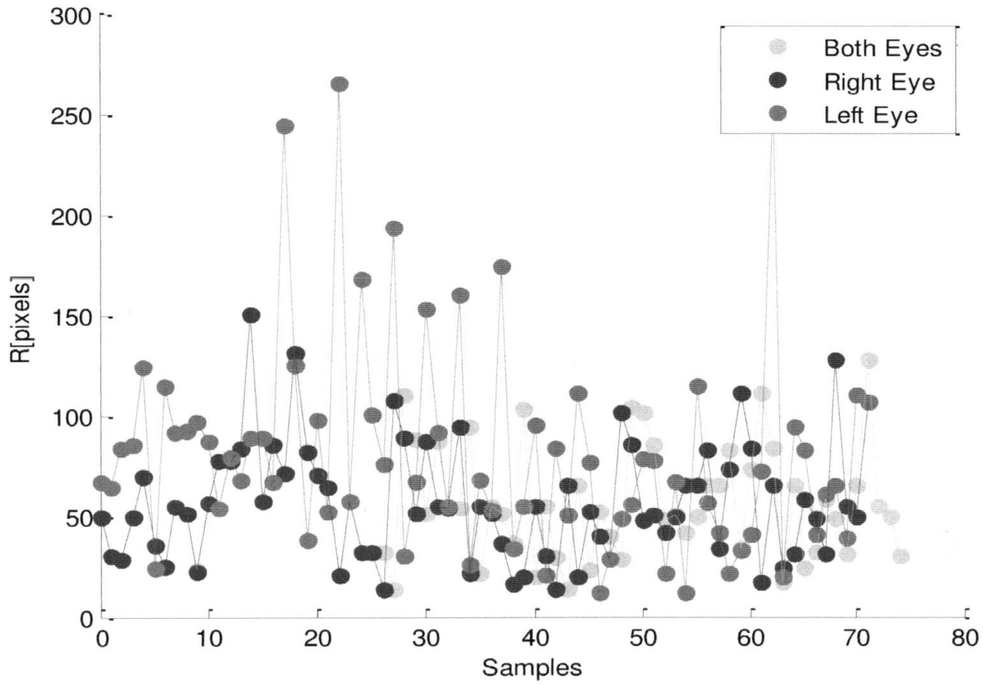

Figure 12: Eyes detected

5.3. Displacement

In this case we have no direct access to this quantity but we can determine it from the pupil position of both eyes. This is an available data, also in real time, and help us to determine whether there was a displacement. We are going to divide the possible displacements in three categories: vertical, horizontal and depth. The vertical displacement is the motion of the head up or down with respect to the reference position.

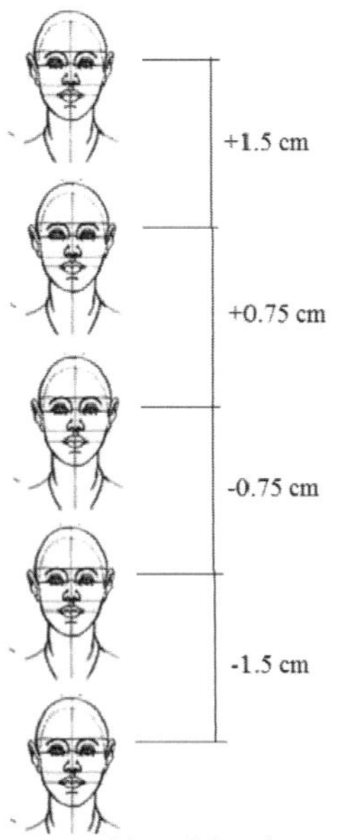

Figure 13: Head positions considered for the vertical displacement

This can be calculated starting from the y position of the pupils, calculated as the mean of the left and the right one. The tests to determine this parameter are done moving the head upward of 0.75 cm and then of 1.5 cm and the same moving it downward. Over this position the eyes are out of the camera field of view and hence it has no sense to do any test. In Figure 13 can be seen the different positions, from the eye tracker point of view, at which the tests were made. In this case we take the data coming from the different tests and interpolate them with two linear functions since the behaviour does not seem symmetric with respect to the center. This can be explained by the fact that the camera used by the eye tracker is in the bottom part of the screen. The data and the interpolation can be seen in Figure 14.

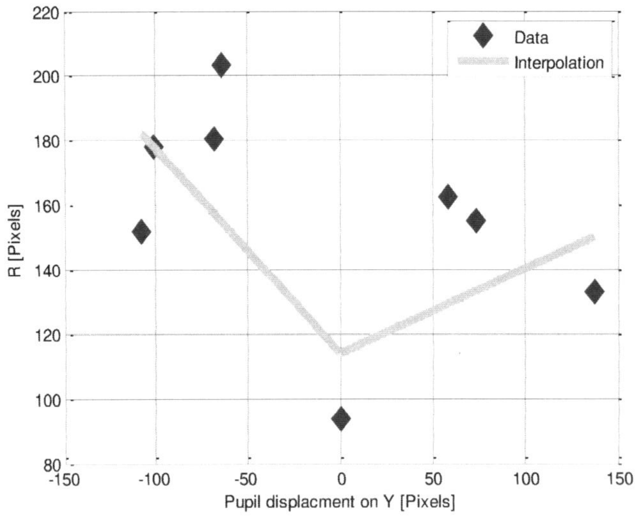

Figure 14: Vertical displacement

On the y axis there is the radius, while on the x axis there is the difference between the reference y pupil position and the current one. Both the quantities are in pixels. Note that the radius is always calculated considering the value that contains the 95% of the available values. The relation we finally obtain is the following:

$$\frac{dR}{dD_a} = \begin{cases} -0.064 \; if \; D_a > D_{a0} \\ \;\; 0.264 \; otherwise \end{cases}$$

where D_{a0} is the reference value of the y pupil position, and D_a is the current one.

Regarding the horizontal motion, we consider a displacement that goes from 2 cm to 6 cm.

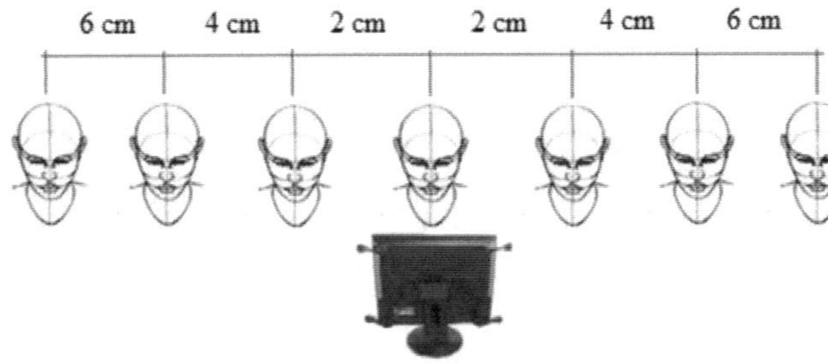

Figure 15: Horizontal head displacement considered

Again an higher displacement will be meaningless since the eye will fall out of the camera field of view. Also in this case we take the data coming from the different tests and interpolate them with two linear functions. Unlike the vertical motion, the behaviour seems symmetric with respect to the center. Hence the two linear function will be the same. The data and the interpolation can be seen in Figure 16.

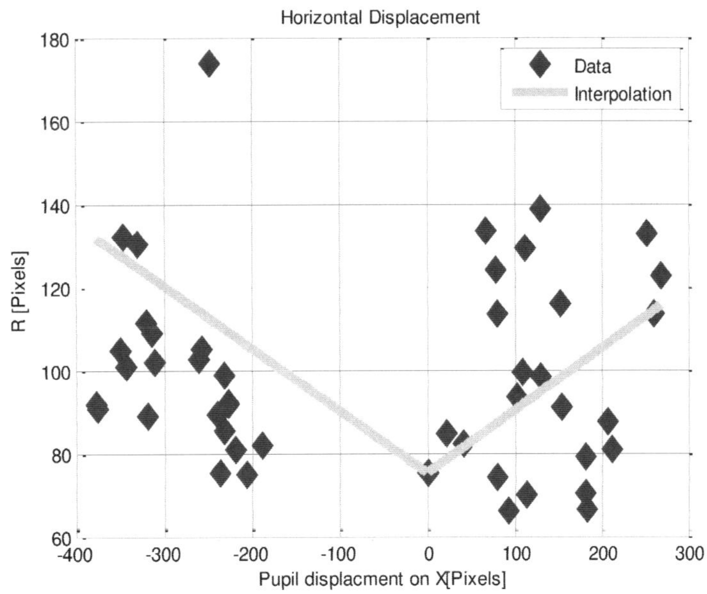

Figure 16: Horizontal Displacement

As before on the y axis there is the radius, while on the x axis there is the difference between the reference y pupil position and the current one. Both the quantities are in pixels. Note that the radius is always calculated considering the value that contains the 95% of the available values. The relation we finally obtain is the following:

$$\frac{dR}{dD_b} = 0.15$$

Finally the last displacement we will consider is the one in the direction perpendicular to the screen. The default measurement sare done standing at about 40 cm from the screen. The tests to determine the influence of the distance are done at a distance of 42.5, 45 and 47.5 cm from the screen.

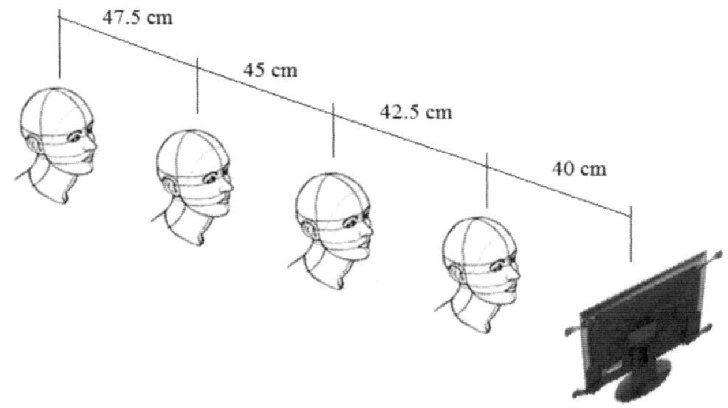

Figure 17: Different depth considered in the test

To understand if the user moves from the reference position we can use the distance between the two pupils. The nearest the two eyes the further the head from the screen. The data obtained and the interpolation can be found in Figure 18. We use a linear interpolation since it seems to fit well the values obtained, with some few exception that are considered as outliers.

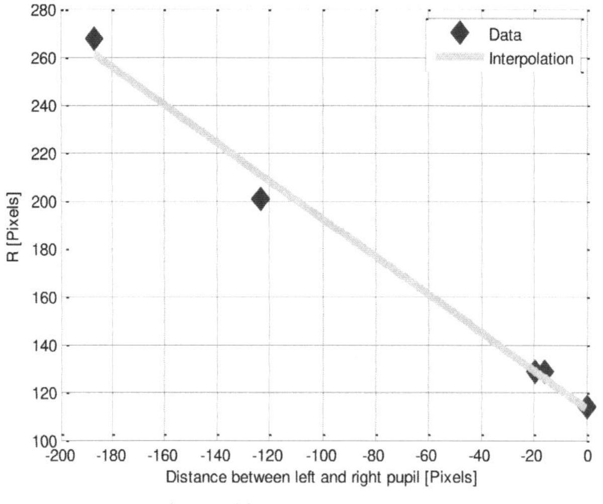

Figure 18: Depth displacement

In this case the model obtained is the following:

$$\frac{dR}{dD_c} = 0.767$$

5.4. Rotation

The last parameter of influence considered is head rotation. In this case we consider the rotation to the left and to the right. We make two cases for both direction verifying that the two eyes remain visible in all the situations considered. Indeed if one of the two eye is not visible any more we will fall in the previously discussed case of the number of eyes. The rotation considered are of 5 and 10 degrees both to the right and to the left.

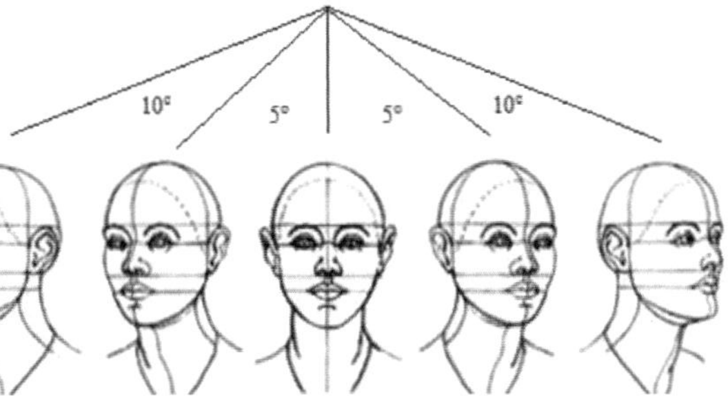

Figure 19: Rotation considered

In order to determine whether a person is rotating is head or not we would like to use the information about the 4 infrared light in the eyes, see Figure 20.

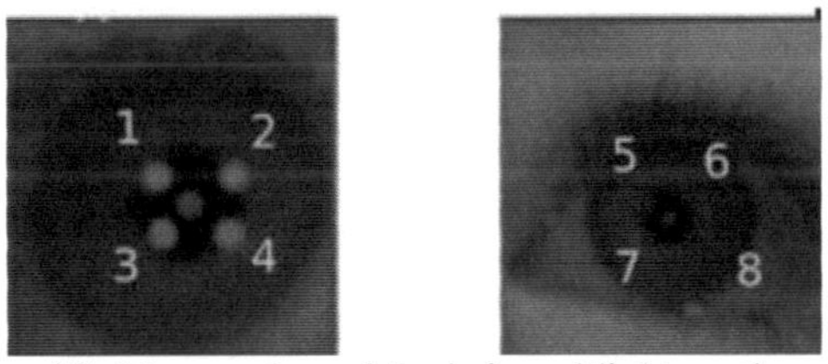

Figure 20: Numeration of the infrared light on the pupils

In order to obtain a value that is proportional to the rotation we take the x position of the point on the right (1 and 3 for the right eye, 5 and 7 for the left eye) minus the x position of the point on the left (2 and 4 for the right eye, 6 and 8 for the left eye). For example for the right eye we get:

$$\Delta x = \frac{(x_1 + x_3)}{2} - \frac{(x_2 + x_4)}{2}$$

To better understand the numeration see Figure 20. This quantity should diminish when rotating. The results obtained are plotted in Figure 21. As explained in the legend the green point is the reference, the blue points are the data coming from the right rotation and eventually the red ones are the data coming from left rotation. On the x axis we have the difference between the distance of the points in the eye, calculated as explained before, and the reference distance, that is the one obtained with the reference test. In the reference test the head is obviously not rotated. On the y axis, as in the previous cases, we have the radius of the covariance circle. From Figure 21 we can see that we have two main issues. The first is that the difference between the pupils positions does not change that much, the maximum is less than 1 pixel. Those kind of displacement are so little that we can't say if they are noise or truly there is a rotation. Indeed some data result even bigger than the reference. The second issue is that the data obtained in this way are very noise and does not seem to follow any particular trend. Indeed in some cases the performances seem to improve with the rotation. This two issues make it not reasonable to model the rotation in our case.

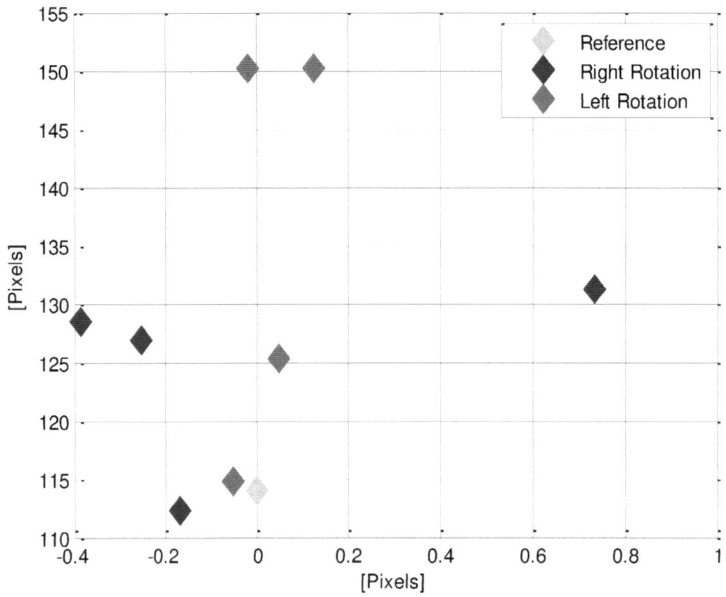

Figure 21: Rotation data

5.5. Validation

Combining the above modelled effects, we achieve the final expression that models the influence of each parameters on the uncertainty radius as follows:

$$R = R(P_0, N_0, D_{a0}, D_{b0}, D_{c0}) + 0.009\Delta P + 71n + d_a\Delta D_a + 0.15\Delta D_b \; + \; 0.767\Delta D_c \qquad \text{[pixels]}$$

Where:

- n it's a variable that determines if we have data coming only from the left eye. Equal to one in this case and zero in the other.
- d_a it's a value that depend on the direction of the vertical displacement. It is equal to -0.064 if the head is moving upward with respect to the reference position and 0.264 in the other case.

It has been implemented a validation test. That based on the incoming data determines a value for the radius R according to the above model. The target in this case is the mouse that was moved along random positions on the screen. The user was asked to move his head during the test. The data can be seen in Figure 22.

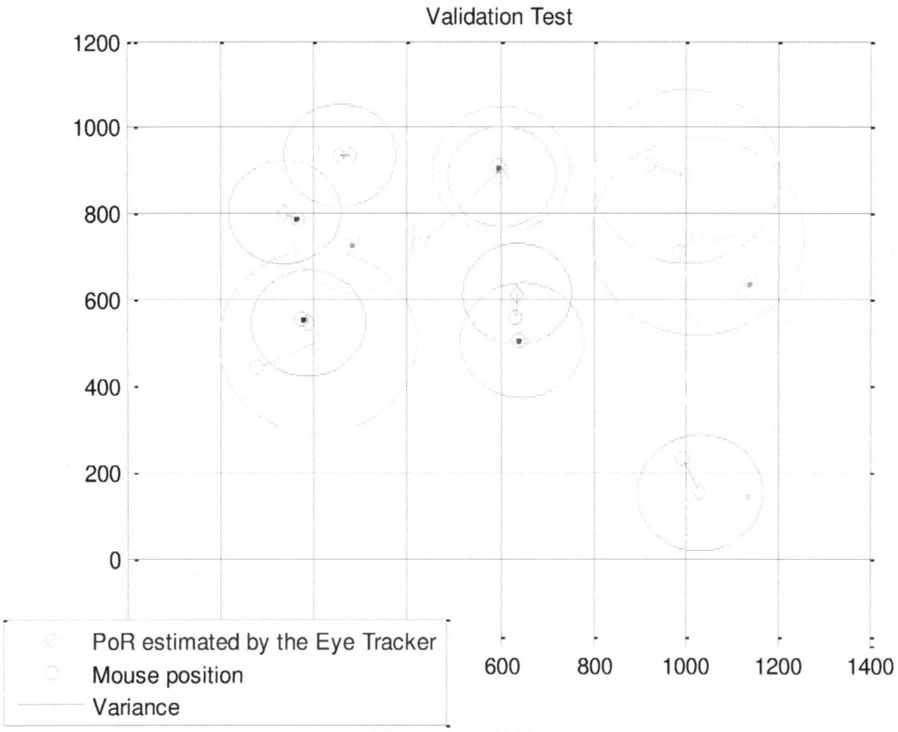

Figure 22: Validation

In the above figure we have 20 data, that are a subset of the data considered for the validation, and only one is outside the estimated uncertainty. Moreover we determine the percentage of target that are inside the equiprobable circle. Doing four tests, each containing 50 data after different calibrations, we obtained 86% of valid data. That is in the 86% of the cases the target fall in the circle.

6. Conclusion

The eye tracker performances have been estimated focusing on its uncertainty as a function of different influencing factors. We have found that the uncertainty can be simplified to a circle, hence depending on only one parameter R. we computed R in nominal conditions taking into account also the systematic error that is indeed random. The value obtained for R in nominal conditions was 114 pixels. Since the screen was 1280x1024 this correspond to the 9% of the screen height and the 11% of the screen width. In addition to the value obtained for nominal operating conditions we estimated the dependency of the uncertainty from a set of influencing parameters. In particular we determined the relationship between the uncertainty and those parameters that can be estimated in real time: target position on the screen, eyes detected, head position and rotation with respect to the instrument. The value obtained for uncertainty was applied to a set of validation data. For the nominal case we get that in the 95% of the cases the circle contains the true value, in the cases affected by the influencing parameters the percentage lowered to 86%. One explanation of the discrepancy could be found in that the influencing parameters were considered as not correlated.

References

[1] Rayner, Keith, *Eye movements in reading and information processing: 20 years of research,* Psychological bulletin 124.3 (1998): 372.

[2] Arne John Glenstrup and Theo Engell-Nielsen, *5 Applicability of Eye-Gaze Tracking Techniques*, http://www.diku.dk/~panic/eyegaze/node23.html

[3] Al-Rahayfeh, A. and Faezipour, M., *Eye Tracking and Head Movement Detection: A State-of-Art Survey*, Translational Engineering in Health and Medicine, IEEE Journal of 2013,1,2100212-2100212, doi10.1109/JTEHM.2013.2289879, ISSN 2168-2372.

[4] Carlos H. Morimoto, Marcio R.M. Mimica, *Eye gaze tracking techniques for interactive applications, Computer Vision and Image Understanding*, Volume 98, Issue 1, April 2005, Pages 4-24, ISSN 1077-3142, http://dx.doi.org/10.1016/j.cviu.2004.07.010. (http://www.sciencedirect.com/science/article/pii/S1077314204001109)

[5] Kenneth Holmqvist, Marcus Nyström, and Fiona Mulvey., *Eye tracker data quality: what it is and how to measure it. In Proceedings of the Symposium on Eye Tracking Research and Applications*, (ETRA '12), Stephen N. Spencer (Ed.). ACM, New York, NY, USA, 45-52. DOI=10.1145/2168556.2168563 http://doi.acm.org/10.1145/2168556.2168563.

[6] Nyström, Marcus and Andersson, Richard and Holmqvist, Kenneth and van de Weijer, Joost, *The influence of calibration method and eye physiology on eyetracking data quality*, Behavior Research Methods,volume 45,1,doi = 10.3758/s13428-012-0247-4, http://dx.doi.org/10.3758/s13428-012-0247-4, Springer-Verlag, pages=272-288.

[7] TOBII TECHNOLOGY (a).2011. *Accuracy and Precision Test Method for remote Eyetrackers*, rev 2.1.1, 7 February 2011, Tobii Technoglogy AB, retrived from http://www.tobii.com

[8] Clemotte A., Velasco M., Torricelli D., Raya R. and Ceres R. (2014). *Accuracy and Precision of the Tobii X2-30 Eye-tracking under Non Ideal Conditions*, In Proceedings of the 2nd International Congress on Neurotechnology, Electronics and Informatics, ISBN 978-989-758-056-7, pages 111-116. DOI: 10.5220/0005094201110116

[9] Dan Witzner Hansen, Arthur E.C. Pece, *Eye tracking in the wild*, Computer Vision and Image Understanding, Volume 98, Issue 1, April 2005, Pages 155-181, ISSN 1077-3142, http://dx.doi.org/10.1016/j.cviu.2004.07.013. (http://www.sciencedirect.com/science/article/pii/S107731420400116X)

[10] White, K.P., Jr.; Hutchinson, T.E.; Carley, J.M., *Spatially dynamic calibration of an eye-tracking system*, In Systems, Man and Cybernetics, IEEE Transactions on , vol.23, no.4, pp.1162-1168, Jul/Aug 1993. doi: 10.1109/21.247897

[11] Juan J. Cerrolaza, Arantxa Villanueva, Maria Villanueva, and Rafael Cabeza. 2012. *Error characterization and compensation in eye tracking systems*, In Proceedings of the Symposium on Eye Tracking Research and Applications (ETRA '12), Stephen N. Spencer (Ed.). ACM, New York, NY, USA, 205-208. DOI=http://dx.doi.org/10.1145/2168556.2168595

[12] Guestrin, E.D.; Eizenman, E., General theory of remote gaze estimation using the pupil center and corneal reflections, In Biomedical Engineering, IEEE Transactions on , vol.53, no.6, pp.1124-1133, June 2006. doi: 10.1109/TBME.2005.863952

[13] Brolly, X.L.C.; Mulligan, J.B., *Implicit Calibration of a Remote Gaze Tracker*, in Computer Vision and Pattern Recognition Workshop, 2004. CVPRW '04. Conference on , vol., no., pp.134-134, 27-02 June 2004, doi: 10.1109/CVPR.2004.92

[14] Sheng-Wen Shih; Yu-Te Wu; Jin Liu, *A calibration-free gaze tracking technique, in Pattern*

Recognition, 2000. Proceedings. 15th International Conference on , vol.4, no., pp.201-204 vol.4, 2000, doi: 10.1109/ICPR.2000.902895

[15] http://www.gametheory.ch/index.jsp?positionId=101139

[16] Randall C. Smith and Peter Cheeseman. 1986. *On the representation and estimation of spatial uncertainty. Int.* J. Rob. Res. 5, 4 (December 1986), 56-68. DOI=10.1177/027836498600500404 http://dx.doi.org/10.1177/027836498600500404

IOP Conf. Series: Journal of Physics: Conf. Series **778** (2017) 012003 doi:10.1088/1742-6596/778/1/012003

Photoelasticity and DIC as optical techniques for monitoring masonry specimens under mechanical loads

C Colla[1] and E Gabrielli[2]

[1,2] DICAM Dept., School of Engineering and Architecture, University of Bologna, v.le Risorgimento 2, 40136 Bologna, Italy

camilla.colla@unibo.it

Abstract. To evaluate the complex behaviour of masonry structures under mechanical loads, numerical models are developed and continuously implemented at diverse scales, whilst, from an experimental viewpoint, laboratory standard mechanical tests are usually carried out by instrumenting the specimens via traditional measuring devices. Extracted values collected in the few points where the tools were installed are assumed to represent the behaviour of the whole specimen but this may be quite optimistic or approximate. Optical monitoring techniques may help in overcoming some of these limitations by providing full-field visualization of mechanical parameters. Photoelasticity and the more recent DIC, employed to monitor masonry columns during compression tests are here presented and a lab case study is compared listing procedures, data acquisitions, advantages and limitations. It is shown that the information recorded by traditional measuring tools must be considered limited to the specific instrumented points. Instead, DIC in particular among the optical techniques, is proving both a very precise global and local picture of the masonry performance, opening new horizons towards a deeper knowledge of this complex construction material. The applicability of an innovative DIC procedure to cultural heritage constructions is also discussed.

1. Introduction and aims

The response of masonry structures under mechanical loads is very complex and not easily determinable mainly due to their heterogeneous character, long life-cycle and effects from the environment. Numerous factors such as characteristics properties of the single component materials, the arrangement of the units, the properties of the interfaces between mortar layers and brick units affect the masonry behaviour, its durability and strength [1]. Masonry being a composite material used since ancient time, largely widespread and with innumerable local variations, several researches have been carried out to evaluate the mechanical behaviour of masonry and of masonry structures, as well as to establish correct ways of analysis. From the computational viewpoint, numerical models are developed and continuously implemented at diverse scales from micro to macro-scales [2], with the aim of obtaining as realistic as possible evaluations. Here, in addition to the theoretical approach adopted, the results largely depend on the required input parameters, which may derive from estimations or from experimental determinations. On the other hand, from an experimental viewpoint approach, standard mechanical tests are available for measurements in the laboratory or, less common, on site. In these cases, instrumentation of the specimen or masonry portion is generally achieved by means of traditional measuring devices, such as linear variable displacement transducers or strain gauges, thus performing punctual and "blind" measures of strains and stresses. These instruments are quite popular due to their low-cost character, their replacement easiness and possible re-use together with the advantages of obtaining a rapid evaluation of principal strains in specific directions.

Nonetheless, drawbacks exist also by this consolidated way of operating: values are collected in few points where the tools are installed and extracted figures are assumed to represent the behaviour of the whole specimen or of large portions of it. This may be optimistic or too approximate in many cases due, for example, to the contemporarily presence of stress concentration areas and stress release areas in masonry under loading. These were documented even for masonry with simple geometries and far from failure loading [3]. Optical monitoring techniques such as photoelasticity or image correlation methods may help overcoming some of these limitations by providing full-field information of the area under test. The production or visualization of test outcome in the form of parameter maps allows qualitative and/or quantitative appreciation of distribution variations. Accurate determination of mechanical parameters may also be possible at local points or over lines and areas. These methods are suitable for almost any kind of material and element subjected to static or dynamic loads [4]; hence, they have the potential to be applied to investigate the real development and distribution of strains in masonry too. Surprisingly, optical methods, although known since decades [3], are still very scarcely employed in civil engineering; even fewer were the experimental attempts of studying the behaviour of masonry structures by optical methods. From a rapid literature search it appears that experimental monitoring of masonry tests via photoelasticity were carried out only on dry masonry models in which the units were made by photoelastic materials [3], [5]. Instead, initial but very promising investigations were carried out via 2D digital image correlation for assessing the physical or mechanical behaviour of masonry [6]-[10].

In this paper, a novel experimental work is presented. It was carried out with the purpose of obtaining an insight into the mechanical behaviour of regular brick masonry when undergoing mechanical compression tests in the lab under eccentric load configurations. Monitoring of two tests over the time was undertaken by two non-destructive optical monitoring methods – photoelasticity and digital image correlation. Testing aims, procedures, data acquisition and data post-processing phases are described for a laboratory study on brick masonry and are compared with traditional, punctual measurement tools. Their advantages and limitations are discussed. Hence, the feasibility and reliability of the two selected optical techniques in providing a deeper knowledge on the response of brick masonry specimens subjected to eccentric compression tests is evaluated. Finally, further developments and applicability of image correlation method for on-site tests on cultural heritage and historic structures are also discussed.

2. Basic principles of the employed optical methods

2.1. *Photoelastic method.*

Photoelasticity is a non-destructive optical technique for the analysis of strain and stress distributions in elastic bodies. It is based upon a unique property -double refraction or birefringence- of some transparent materials such as some plastics, crystals and polymers. This is proportional to the strain and stress states of the element, artificially transferred in photoelasticity. It has been proved that, a loaded photoelastic specimen in combination with optical elements and a source of illumination, exhibits fringe patterns related to the principal stress differences in a plane normal to the light propagation direction. This technique has been mainly used for two dimensional plane problems, although, thanks to recent developments, tri-dimensional analyses are now possible too [11], [12], [13]. The photo-elastic coating method is an extension of photoelasticity that allows studying also inelastic bodies. It has been used for a variety of applications ranging from automotive, to aircraft and aerospace, to biomechanical and composite materials [13]. It is based on the direct bonding by adhesive of a thin layer of a photoelastic material (commercially available in flat-sheet or liquid form) on the surface of the element to be tested. The photoelastic effect is studied via a reflection polariscope, which illuminates the monitored surface via a polarized light. It is assumed that the strains in the structure under load and in the superficial coating will be the same. The strains in the coating layer produce proportional optical effects visible as isochromatic fringes, which are series of successive and contiguous different-coloured bands, each one representing a birefringerence degree, thus, corresponding to a specific strain level. Moreover, with a reflex polariscope, by rotating the planes of polarization of the incident light, isoclines lines may also be isolated. These connect the points where the directions of principal strains are parallel to a given direction. Hence, with this optic method it is possible to distinguish between highly stressed areas from low stressed areas i.e. where structural material is not efficiently used thus carrying out quantitative analyses [12], [13].

2.2. Digital image correlation technique.

Digital image correlation (DIC) is a quite recent non-destructive, optic, contactless method which exploits the principles of photogrammetry, digital image processing and stereo-images correlation to determine high definition displacement and strain maps of an object subjected to external forces [4]. Generated in the early '80s as a 2D method to measure the in-plane displacement field of an object while under load, more recently the DIC concept was extended to stereovision systems. Hence, a 3D version of the method was developed. This is based on the simultaneous acquisition of pairs of images, by means of two CCD cameras placed in front of the investigated surface in a stereoscopic configuration (Figure 1 left) thus allowing measuring also the out-of-plane motion [4], [14]. During data acquisition, pairs of images are collected at diverse deformation states of the investigated element. Then, each couple of images with deformed geometry of the element under tests is compared to the pair of reference images recorded in the specimen's undisturbed configuration. A stereo imaging equation is applied followed by a correlation algorithm performing a pixel-by-pixel comparison of images. That is the same grey level of each pixel is tracked at diverse time instants of the test (Figure 1 right). Thanks to a DIC calibration procedure to be performed prior to the test start, the parameters characterizing the specific adopted optical measurement configuration are registered (in terms of distortion of the lenses, relative distance of the cameras, orientation of the cameras with respect to the object surface, ...) [4]. These values are used to calculate the reference axis system and establishing the ideal working plane. *Thus, a*t the end of the data analysis, the 3D coordinates (x, y, z) of each point P of the images are accurately determined. From these values, the 3D displacement vector [u(P), v(P), w(P)] and corresponding deformations fields [ε_{xx}, ε_{yy}, γ_{xy}] during the whole test duration are calculated. DIC results are usually plotted as coloured maps of each geometrical or mechanical parameter superimposed to the digital photos [15] nevertheless parameter values can be extracted along segments in addition to single points or areas.

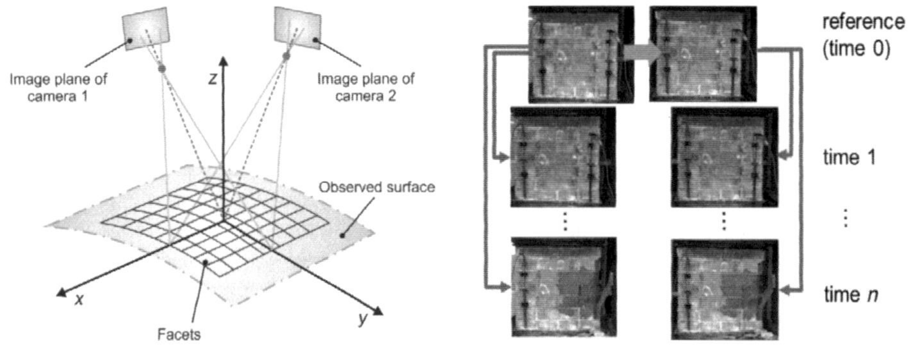

Figure 1. Sketch with principles of 3D DIC in stereoscopic configuration of the cameras (left) [16]; image comparison for data analysis procedure (right).

3. Experimental work

3.1. Description of the masonry specimens and mechanical tests

Two brick columns were studied, each made of 6 clay bricks and interposed hydraulic lime mortar joint. The columns had nominal dimensions of 250 x 120 x 380 mm^3 (length x width x height) while the units – new fired-clay bricks - had dimensions of 250 x 120 x 55 mm^3 (length x width x height);

the mortar joints were nominally 1 cm thick. The main mechanical and physical properties of the masonry component materials were determined via standard tests at the time of building the specimens (Table 1) with the purpose of estimating the strength of the specific masonry under construction. After appropriate curing period of the specimens to allow for the strength development in the mortar, the columns were prepared and instrumented for the mechanical tests. In order to simulate a realistic stress-strain configuration as it may be found in existing masonry structures, an eccentric load compression test was preferred to a standard configuration with uniform distribution of vertical in-plane load [17], [18] commonly representing the laboratory set-up for assessment of masonry mechanical properties. For these tests, the testing machine was equipped with a metal rig including the double cylindrical hinges for exerting the eccentric load [19], [20].

The load was applied with sufficient eccentricity (80 mm) from the middle vertical axis of the specimen, for it to fall outside the middle third of the central core of inertia of the specimen section (e/D>1/6 where e=eccentricity and D=base length of the specimen). See Figure 5 left.

Firstly, the specimens were instrumented via traditional or established displacements measurement tools: linear variable differential transducers (LVDTs) on the front and rear faces of the specimens in addition to TML displacement transducers on the lateral faces. Three LVDTs HBM WA series with 0-10mm range and ±1% accuracy were installed on the front side of each column, with vertical alignment to monitor the specimen's global vertical displacements; additionally, on the specimen monitored by DIC, a fourth LVDT was installed at mid- section of the rear face of the column (Figure 2). Six TML UB-5 displacement transducers (also called "omega", due to their shape) were installed on the lateral faces across the 2nd, 3rd and 4th mortar joints (numbering from the base), to monitor their opening/closing movements (Figure 2). These are able to record displacements up to 5 mm with a sensitivity of 1000 microepsilon/mm. This mortar joint monitoring in masonry via omegas is an innovative use of these tools, as they are more commonly applied for monitoring crack opening displacement in metal or in concrete element bending tests. All these devices, as well as the testing machine, were linked to the acquisition system through two 8-channel units. Load and displacement values were registered throughout the test and stored in a PC equipped with dedicated Labview software for both real time visualization and data post-processing. The installation of the supports for anchoring the transducers was quite a time consuming step because it requires precision in positioning and gluing on the irregular masonry surface. The presence of superficial roughness, dust and also the cables' weight may force to repeat this preparatory operation in case of support detachment. Due to these difficulties and in order to overcome the limitations connected with the punctual character of the obtainable values, it was planned to monitor the tests also by innovative non-contact, full-field methods. The two selected methodologies – photoelasticity and digital image correlation – are described in the following text sections.

The two mechanical tests were performed at the LISG laboratory of Bologna University via a Metrocom 600 kN universal testing machine in displacement control mode. According to [21] the expected collapse stress level of masonry to be reached during the compression test was estimated from the characteristic mechanical properties of the single masonry materials to be approximately 15 MPa. The load application velocity was set to about 0.5 mm/min. Each test consisted in loading/unloading cycles at progressively increasing load levels up to the fragile collapse of the brick columns.

Table 1. Average physical and mechanical properties of masonry construction materials.

Material	Bulk density [kg/dm^3]	Porosity [Vol.-%]	Compressive strength f [N/mm^2]	Bending strength f [N/mm^2]	E [GPa]
Brick	1.79	23.2	80	/	16
NHL mortar	2.06	20.6	3.2	1.2	4.8

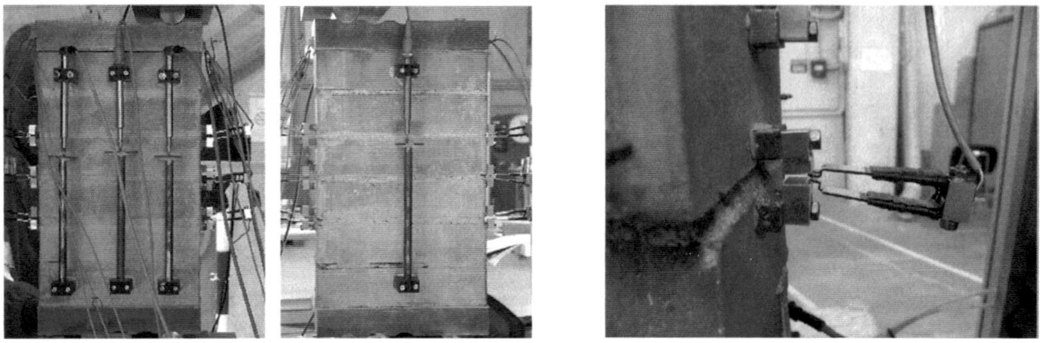

Figure 2. Six-brick column 0S4: position of vertical LVDTs on the front (left) and rear face (centre) of the specimen; detail of a TML transducer installed across a mortar joint on a lateral face (right).

3.2. Experimental monitoring via photoelasticity

In this experiment, optical monitoring was carried out via photoelastic coating method. According to the photoelastic testing facilities present in the laboratory, the aim was to obtain at least a non-destructive qualitative visualization of the actual stress/strain distribution in the masonry specimen while subjected to eccentric vertical loading.

Firstly, a photoelastic coating material with coherent adhesive supporting material needed to be selected. The selection would be a compromise between the desired resulting reliability and accuracy, the sensitivity of the polariscope system to be employed, adhesive setting time and photoelastic material costs. Taking into account the expected masonry compression stress level, the inelastic behaviour of masonry and assuming a flat masonry surface, it was chosen a VishayMM PS-3C photoelastic flat-sheet material with area dimensions 10''x10'' (254x254 mm^2) usually employed for analysis on non-metallic materials. This is a medium-modulus, high-elongation coating sheet, just less than 1 mm thick, characterized by a K factor (strain optical coefficient) of 0.02. The sheet was glued onto the rear face of one of the two masonry columns (named 0S2) by a uniform layer of VishayMM PC-6 epoxy resin. This is a room-temperature-curing, low-modulus, high-elongation adhesive (50% elongation). It requires at least 24 hours setting at temperatures between 21 and 24°C. Although in reality the masonry specimen surface is not completely flat, especially at the edges of the bricks and at the interfaces with mortar joints, a photoelastic coating flat-sheet was preferred to a liquid coating material whose spread on the masonry surface would also have been affected by the irregularities of masonry due to the pores and superficial roughness of the bricks and mortar joints, hence it would have been difficult to mould. Several critical issues emerged in this preparatory phase given that the coating application on the rough, porous and irregular masonry surface was difficult and time-consuming. Despite the expert operator's efforts in accurately cleaning the masonry surface to remove traces of dust and dirt before applying the resin, in filling all the open pores and thus in creating a smooth and regular adhesive surface, it remained in question if a completely uniform thickness of the resin was laid and if this layer was as plane and thin as planned. These aspects may affect the perfect adherence of the photoelastic sheet to the specimen's surface and the response of the photoelastic material to the stress and strain transfer from the masonry.

With reference to the photoelastic material itself, pre-manufactured sheets are more economic and easier to use than liquid coatings and they guarantee uniform thickness and uniform physical, mechanical and photoelastic properties [22]. The predefined dimensions of the commercial sheets may be a limitation for specific applications. In the specific case, the sheet area was as large as the brick length but not long enough to cover the whole masonry height. Therefore, it was positioned to cover the lower masonry area with the first 4 masonry joints leaving unmonitored the top brick in the column (Figure 3 left). The whole test was monitored via a conventional reflection polariscope equipped with a high-intensity white light source (Luchsinger 030-series, 50 Hertz), together with a medium-resolution digital video camera JVC. In order to record the whole coated area, the system was mounted on a tripod and positioned at about 90 cm distance in front of the specimen (Figure 3 centre and right).

Figure 3. Surface preparation: pouring of the resin (top left) and application of photo-elastic sheet (bottom left); test set-up (centre) and monitoring via reflective polariscope (right).

3.3. *DIC monitoring of mechanical tests*

The second optical monitoring technique considered was digital image correlation (DIC) in 3D configuration. A commercial system (VIC-3D) available at LISG laboratory is equipped with in two digital cameras (5MPx), with 23 mm Schneider lenses, a 8-channel data acquisition/synchronization unit and a laptop with dedicated software. The system is able to determine deformations with accuracy up to 50 $\mu\varepsilon$. During the test, the two cameras were fixed on the same arm and rigid tripod at a relative distance of 27 cm and at 1.45 m away from the investigated masonry surface. A white lamp with diffused light was placed between the cameras and the tested masonry surface to minimize the non-uniform brightness field caused by the scarce ambient illumination of the laboratory; moreover, a sunshade was used to protect the specimen surface from sunlight changes, which would vitiate the method performance (Figure 4 left).

Figure 4. Six-brick column 0S4: set-up of mechanical test monitored by DIC (left); post-processing selection of analysis area (centre); positions of virtual extensometers (right).

The test was carried out adopting an innovative testing procedure developed by the authors, which consists in employing the natural masonry texture as reference pattern for the DIC readings. This set-up overcomes the preparation of the investigated surface, usually a technique's mandatory requirement, performed by uniformly spraying the surface by white paint coating and overlaying a random pattern of appropriate-size black dots [15]. This simplification was developed for an intended

IOP Conf. Series: Journal of Physics: Conf. Series **778** (2017) 012003 doi:10.1088/1742-6596/778/1/012003

further use of DIC on site, in the field of Cultural Heritage or historic structures, where coating would not be allowed [23]. After positioning the DIC equipment in front of the surface to be investigated, the exposure time, focus and aperture range of the lenses were adjusted and the calibration performed obtaining a very small error value (approx. 0.03 pixels) which indicated that the selected monitoring configuration was highly precise despite the adopted simplification. One pixel of the recorded image corresponded to 0.2 mm in reality. Continuous monitoring was first by image acquisition every 5 s whilst every 2 seconds after appearance of the 1st crack. This method is easy-to use and the very fast and simple preparatory phase, does not require high expertise operators. A disadvantage of the method is that high capacity storage memories are necessary.

4. Experimental results and discussion

4.1. Mechanical tests: data analysis and results

During each test, load values were collected together with global and local displacements. From these data, load-displacement and stress-strain curves were plotted and the main mechanical parameters characterizing the behaviour of the masonry columns were calculated.

Being the stress distribution in the specimens' section unknown, the compressive strength assessment was carried out following two diverse methods: 1) assuming a linear strain distribution over the section and considering a simplified perfectly brittle constitutive model; 2) by considering also the tensile contribution of masonry and employing a stress distribution for a partialized resistant cross-section (Figure 5 centre). It is known that the first hypothesis is too limitative to actually represent the masonry behaviour as it does not take into account the inelastic compressive response of masonry and it may led to some overestimations [18], i.e. apparent increase in the compressive strength due to load eccentricity, but it was used just to perform a first discrimination.

Hence, the equations used to calculate the compressive strength values, according to the 1st and 2nd model respectively, are:

$$f_c^{'} = \frac{2N}{3su} \tag{1}$$

$$f_c^{pp} = \frac{N}{A} \pm \frac{Ne}{I} y \tag{2}$$

where $f_c^{'}$ [MPa] is the maximum compressive strength in the external border of the section closest to the load application point; f_c^{pp} [MPa] is the compressive strength in correspondence of the load application point; $f_t^{'}$ [MPa] is the tensile strength; N is the maximum load, e is the load eccentricity; I is the moment of inertia of the section and y is the distance between the barycentre of the section and the load application point.

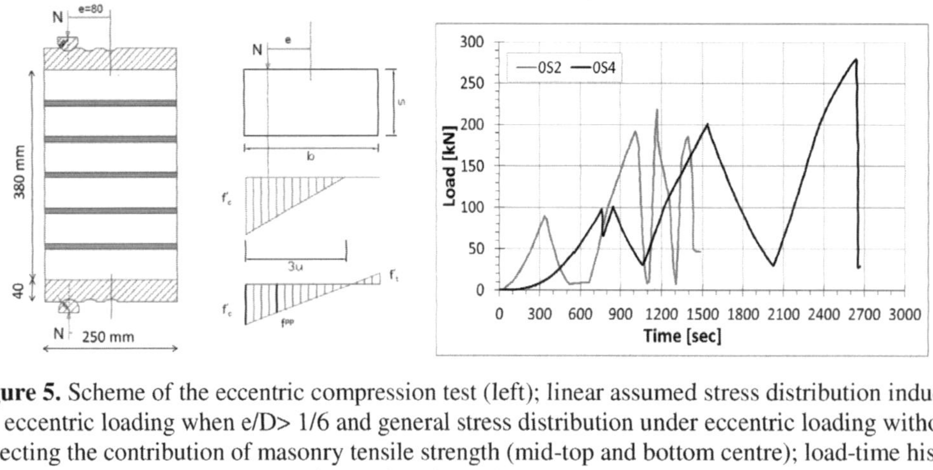

Figure 5. Scheme of the eccentric compression test (left); linear assumed stress distribution induced by eccentric loading when e/D> 1/6 and general stress distribution under eccentric loading without neglecting the contribution of masonry tensile strength (mid-top and bottom centre); load-time history of both six-brick columns (right).

The two mechanical tests lasted between 25 and 45 minutes (Figure 5 right). The maximum compressive loads were equal to 219 kN for six-brick column 0S2, monitored via photoelastic coating method, and almost 280 kN for six-brick column 0S4, monitored via DIC. The displacement values measured via LVDTs were plotted against the applied loads (Figure 6) and then used to calculate the corresponding values of stress and strain as above explained (Figure 6 right). Note that for the six-brick column 0S4 it was not possible to record the post peak behaviour because of the falling of all wired measuring instruments just after reaching the maximum load.

In Table 2 the resulting main mechanical parameters are reported for both masonry specimens and the differences between the values obtained in the two cases are highlighted.

Being the compression load applied with a certain eccentricity the LVDTs placed on the same face at the edges of the specimen, act in opposite directions: compressed under the load application point, in a tensile stress state the other (Figure 6 left). The asymmetrical behaviour (with reference to the vertical axis of the column) was further highlighted by the omegas across the mortar joints, for both columns. A closure tendency was recorded for the three mortar joints of the right side, nearest to the load application points, whilst the other three are opening (see i.e. Figure 10 right).

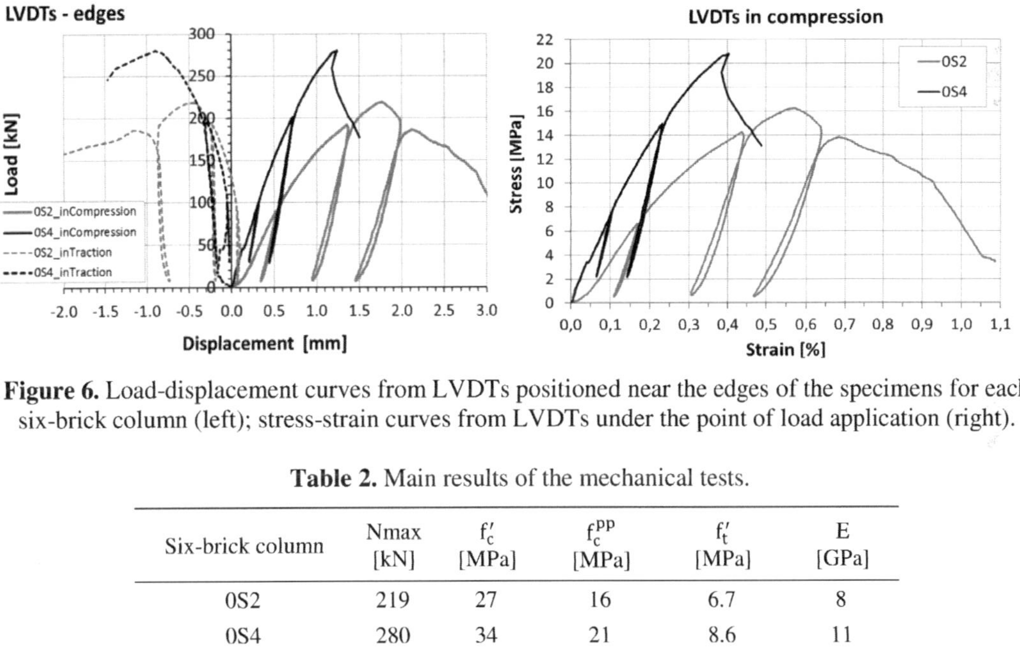

Figure 6. Load-displacement curves from LVDTs positioned near the edges of the specimens for each six-brick column (left); stress-strain curves from LVDTs under the point of load application (right).

Table 2. Main results of the mechanical tests.

Six-brick column	Nmax [kN]	f'_c [MPa]	f_c^{pp} [MPa]	f'_t [MPa]	E [GPa]
0S2	219	27	16	6.7	8
0S4	280	34	21	8.6	11
Difference [%]	*27.9*	*25.6*	*29.4*	*29.9*	*37*

The maximum compressive strength determined according to eq. 2, was of about 16MPa for column 0S2 and 21 MPa for specimen 0S4. The latter specimen is about 30% more resistant than 0S2; this difference clearly shows the inhomogeneous character of masonry structure, also when built with the same construction materials and tested in the lab, in similar conditions. Moreover, at the peak stress value, a compressive strain of 0.57% was measured for specimen 0S2, whilst only 0.40% for specimen 0S4 (Figure 6 right), hence, 0S4 appeared 30% less deformable than 0S2. The values of the secant elastic modulus E calculated from the stress-strain slopes between 0.1 and 0.4 of the maximum compressive strength (f_c^{pp}) underlined the diverse behaviour of the two masonry columns being 0S4 stiffer than 0S2 (about 37%) (Table 2).

Additional differences were found in the failure modes of the two brick columns, although in both cases, the collapse started with the formation of a hinge located at the 4th mortar joint, under the load application point. For six-brick column 0S2, failure continued with a separation plane in correspondence of the 5th mortar joint whilst, for specimen 0S4, two separation planes occurred, involving both the 3rd and 4th mortar joints (Figure 7). These differences are again an index of the irregular behaviour of masonry, making it difficult to predict the reactions of masonry under loads.

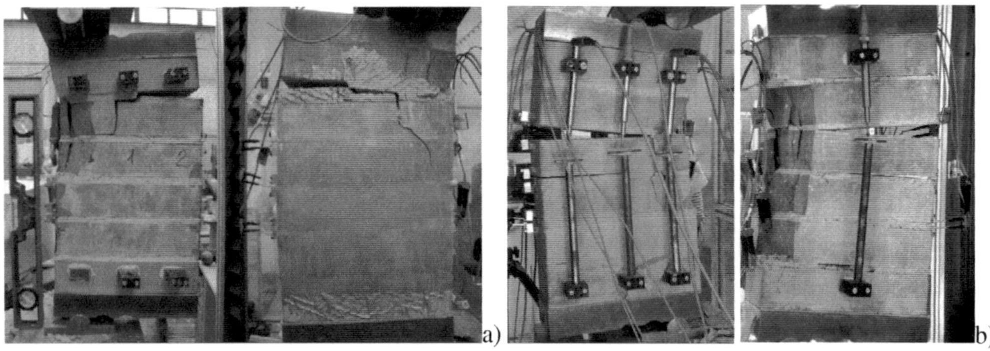

a) b)

Figure 7. Failure mode of a) six-brick column 0S2; b) 0S4. Front and rear views, from left to right.

4.2. Results of photoelastic monitoring: six-brick column 0S2

Throughout the mechanical test and since the initial phases of load application, despite quite low resolution images collected with the old video camera at our disposal and a not-perfectly adherent coating photoelastic sheet, areas of strain/stress concentrations are clearly distinguishable from unloaded areas.

Since the beginning of the test to the 4th and last loading cycle, all the strains appeared concentrated in the mortar joints and at interfaces, whilst the bricks did not present any areas of stress concentration (Figure 8 left). The different behaviour of the two masonry materials was expected due to the lower strength of the hydraulic mortar joints compared to the brick units but herein, for the first time, it was visualized on a real masonry specimen. The stresses are not uniformly distributed along the mortar joints but higher concentrations are visible towards the exterior borders of the same; they increased at increasing load levels while expanding towards the mid-section of the column. Mortar joints interact all over their lengths but the distribution of strain is not linear; hence, it appears not sufficient to measure the movements at the two extremities of the joints (as performed by the installed omega) to determine the behaviour of the entire mortar joint.

From the video collected throughout the mechanical test, it is also clear that the four monitored mortar joints – covered by photoelastic sheet- did not present the same level of strains: highest intensity in the 3rd joint and smallest intensity in the 4th one (again, numbering the masonry joints from the bottom of the column). This difference is very evident at the beginning of the test and tended to decrease during the load application. The difference in sign between the areas of stress concentration on the left hand side of the specimen (traction) and on the right hand side (compression) is clear during the 3rd loading cycle (Figure 8 left). The formation of a hinge at the 4th mortar joint, which caused the failure of the six-brick column, is clearly detectable (Figure 8 center).

It is also interesting to note that the brick units seemed to participate to the stress distributions only after the formation of vertical cracks; in this case, the openings of cracks occurred at the 4th loading cycle (last cycle, see Figure 5 right). These cracks were located on the 4th and 5th brick units (i.e. Figure 8, centre, at approx. the maximum load). The evolution of cracks was monitored during the whole test, up to the fragile collapse of the specimen (Figure 8, right). The collapse of this six-brick column occurred with the separation of the 5th mortar joint, which, unfortunately was outside the coated area, thus, it was not possible to follow the crack formation along this joint.

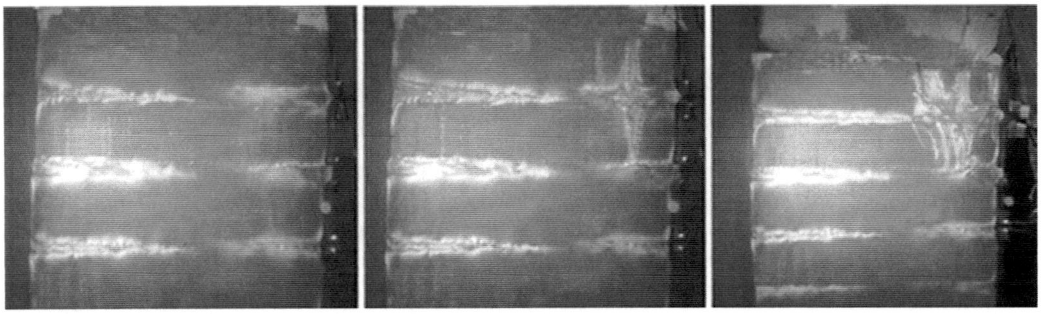

Figure 8. Sequence of images from reflex polariscope at increasing load levels: in the 3rd loading cycle (~120 kN); at maximum load (220 kN); at the end of the test (from left to right).

From this preliminary test, it is clear that the behaviour of masonry under eccentric loading is quite complex. Measuring displacements at the extremities of the mortar joints is necessary but not sufficient to determine the development/evolution of strains along the joints' lengths. Moreover, traditional tools are not able to collect any information regarding crack openings and developments or hinge formations. Thus, it would be desirable to have at disposal a reliable and accurate technique able to give a picture of the whole masonry surface during the entire mechanical test. Photoelasticity has all the potentialities to fulfil this aim at least from a qualitative viewpoint. Quantitative monitoring of strain distributions may also be possible, although herein it was not considered. However, its reliability and accuracy is strongly affected by the difficulties related to the surface preparation. More, the spreading of the adoption of photoelasticity in the civil engineering sector is limited by the complexity of data acquisition and elaboration phases, which require the presence of operators with a certain expertise, as well as by the high costs of the coating materials and by the large amount of waste materials.

4.3. Results of DIC monitoring: six-brick column 0S4

At the end of the mechanical test on the 0S4 specimen, DIC images were easily elaborated via dedicated software [24] by following these steps: 1) calibration recall; 2) selection of the area of interest, to be analysed (Figure 4 centre) and starting points positioning to force the data correlation and not to lose data around any out-of-plane areas (i.e. LVDTs); 3) selection of analysis parameters (such as subset size, step size and noise level, which influences the accuracy of the results); 4) data analysis; 5) rigid motion removal; 6) visualization of results and data interpretation [15]. As already said, at the end of the analysis, data in terms of coordinates, displacements and corresponding strains are available for each correlated pixel of the masonry surface throughout the whole test. However, data are lost when and where the surface is missing (i.e. due to material spalling, cracking, etc.), from that time instant onwards.

Herein, by way of example, DIC results are visualized as 3D coloured maps of strain values ε_{yy} in vertical direction at increasing load levels, superimposed to the deformed images, according to a user's defined colour legend [15], thus obtaining an immediate view of the mechanical response of the six-brick column under eccentric load (Figure 9).

These maps show that at the beginning of the test, there is a null strain state but since very low load levels, i.e. at the first loading cycle (100 kN) areas of strain concentrations become visible at some mortar joints: compressive strains in the 1st mortar joint under load application point; traction state on the 3rd and 4th mortar joints, right hand side (Figure 9left). Strain values are greater on mortar joints than on brick units being the mortar a less resistant and much more deformable material, as already seen also with photoelasticity. However, with this method, the differences are easily quantified. More, the brick units did not appear completely unloaded as some small, low-intensity areas of strain concentration are visible. Another aspect arising from these maps is that the states of strain concentrations are asymmetrically distributed with respect to the middle vertical axis of the specimen, as expected, being the compression load applied with an eccentricity. On the left hand side of the six-brick column, corresponding to the side of load application, there is a state of compression (violet-blue) which, at 100 kN is limited to the 1st mortar joint, whilst on the right hand side of the specimen there is a state of tension (orange-red) that at the same load level, is located on the 4th and 5th mortar joints. This asymmetric behaviour is accentuated with increasing applied load involving a greater number of mortar joints up to all, at the end of the test, with the exception of the 1st joint, which remained compressed until the end of the test all over its length. Both the states of tension and compression increased at increasing loads but of different quantities: the compression state is much greater, almost 3 times, than the tensile state. The formation of rotational hinge on the left side of the 4th mortar joint as well as of the separation planes (horizontal cracks) on the right hand side of the 3rd and 4th mortar joints are well visible with this representation. At the end of test, some parts of the surface were no longer correlated by DIC due to the falling of brick and mortar portions as well as minor detachments occurred all over the specimen surface (grey areas in Figure 9 right). Anyhow, the possibility of collecting information (although partial) also after reaching the fragile collapse of the specimen represents an added value of the technique if compared to all the other traditional, wired measurement tools, as most of them were already disconnected.

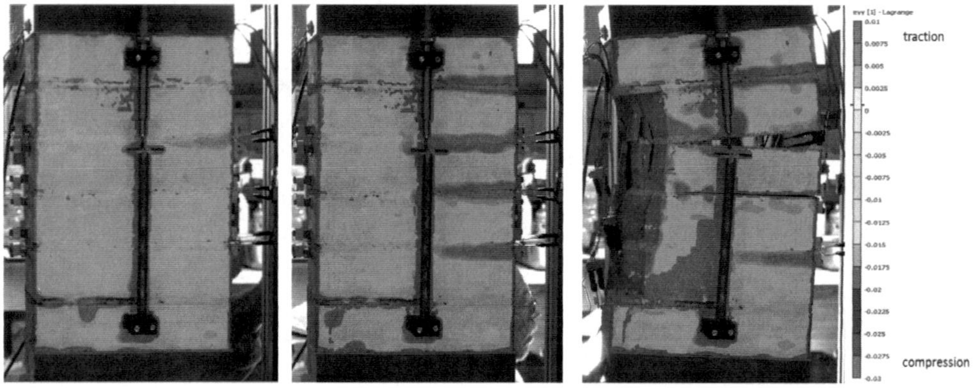

Figure 9. Six-brick column 0S4: Trend of full-field maps of principal strain, ε_{yy}, along the vertical direction at increasing load levels: at the maximum of the 1st and 4th loading cycles and at the end of the test (from left to right), with the same colour legend [$-0.03<\varepsilon_{yy}<0.01$].

Finally, to evaluate the accuracy of this optical method, a comparison was performed among data obtained by LVDTs and virtual extensometers which are a tool of the DIC software used to determine the strains between two selected points. Herein, by way of example, the stress-strain diagrams for the two LVDTs (in compression and traction) positioned on the edges of the front side of the specimen are compared with similar curves obtained by virtual tools positioned in the same positions but on the rear side of the specimen (Figure 10 left). The curves present a very similar slope, being coincident in traction. The vertical compressive strain value measured at the maximum stress is about 35% greater for the virtual extensometer whilst, the tensile strain value measured by DIC is only 2% greater than by LVDT (Table 3). The differences may be explained with the different positions of the instruments (specimen's rear face for DIC, front face for LVTDs).

A similar comparison was carried out also considering the opening/closing movements of the mortar joints measured by wired omega vs. DIC virtual tools. Also in these cases, the slopes of the curves are very similar and differences explainable with diverse positions of the instruments.

This brief example of DIC monitoring of masonry under eccentric loads shows several advantages of this technique both compared with traditional wired tools and with photoelastic-coating method. The 3D DIC technique was easier to use (both the set-up and data analysis phase), did not necessarily require a surface preparation, thus being completely non-destructive and allowed obtaining a complete picture of the whole monitored area by accurately performing a tri-dimensional measure of the geometry of every point of the surface during the whole test duration. Thus, gaining an insight into the mechanical response of masonry; this should help in obtaining a better understanding of the complex behaviour of this composite material. Moreover, there are no additional costs to those of the testing equipment and there are no waste materials (as, instead, in the case of photo-elastic coating). A disadvantage is given by the high computational requirements in terms of time and space, which are necessary to perform data analysis after long-duration tests. The developed testing procedure, which allows overcoming the need for surface preparation represent a great advantage, especially in view of on-site testing [23], also in case of historic masonry structure.

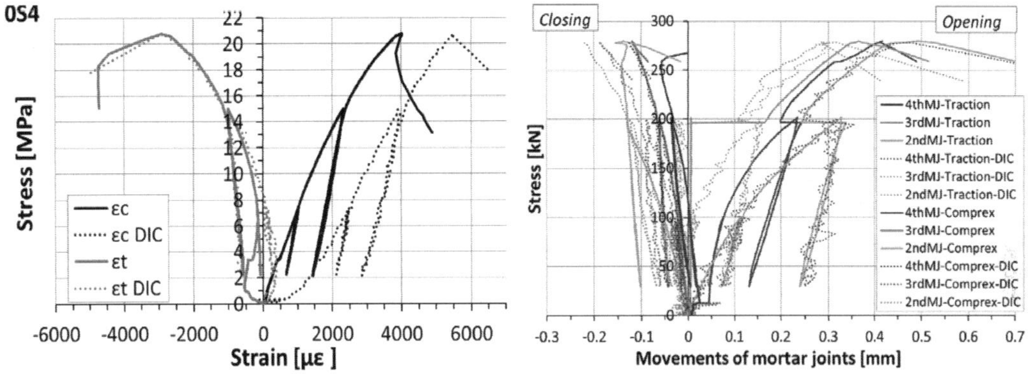

Figure 10. Six-brick column 0S4: stress-strain diagram via LVDTs and virtual DIC extensometers (left); mortar joints movements measured via TML transducers and DIC (dotted lines) (right).

Table 3. Six-brick column 0S4: results from wired and virtual (DIC) measurement tools.

	ε'_c [$\mu\varepsilon$]	ε'_t [$\mu\varepsilon$]	E [GPa]	Avg Closing [mm]	Avg Opening [mm]
LVDT/TML	4027	-2906	11	-0.13*	0.43
DIC	5432	-2962	8.2	-0.18	0.36
Difference [%]	*34.9*	*1.9*	*-25.5*	*39.9*	*-16.1*

* data from two over three TML transducers

5. Conclusions

An experimental laboratory work aimed at assessing the mechanical behaviour of masonry was carried out considering two six-brick columns subjected to eccentric compression tests. Traditional wired tools were used for measuring the displacements during the mechanical tests. The novelty of the work is represented by the additional use of two optical techniques to monitor the mechanical tests. The feasibility and reliability of these methods, still scarcely employed in civil engineering, in obtaining a full-field visualization of mechanical parameters and a better understanding of the behaviour of masonry under loads was tested.

The photo-elastic coating method proved to be suitable for the testing aims, at least for obtaining a qualitative detection of strain concentrations' areas and cracks formation. The global and local behaviour of masonry was qualitatively assessed, obtaining additional information in comparison with the results from punctual measurement tools. Anyhow, the difficulties encountered while preparing the surface to be monitored with a layer of coating material together with the high level of expertise required for a quantitative analysis, the high costs and amount of waste materials are discouraging factors in the spreading of the technique. Moreover, this method would not be feasible for large areas or outdoors masonry because too expensive and invasive.

Instead, digital image correlation (DIC) offers easiness of use and of data elaboration that may allow fast spreading of its adoption also in the construction sector. The full-field visualization of strains herein easily obtained with DIC allows understanding qualitatively and quantitatively the global and local response of masonry under eccentric loads. The differences between masonry component materials were also measured, thus opening new horizons towards a deeper knowledge of this complex construction material. The innovative procedure presented proved to be suitable and accurate, hence opening possibilities for a future application of DIC on-site for monitoring large areas, Cultural Heritage or historic structures.

References

[1] Gabrielli E and Colla C 2015 Experimental insight in the mechanical behaviour of aged masonry specimens *Proc. of AIMETA 2015* 14th -17th Sept. 2015 Genova Italy pp 10

[2] Lourenço PB and Pina-Henriquez J 2006 Validation of analytical and continuum numerical methods for estimating the compressive strength of masonry *Comp. Struct.* **84** 1977-89

[3] Bigoni D and Noselli G 2010 Localized stress percolation through dry masonry walls Part I - Experiments *Eur. J of Mechanics & Solids* **29** 291-298

[4] Sutton MA, Orteu JJ and Schreier HW 2009 *Image Correlation for Shape, Motion and Deformation Measurements*: *Basic Concepts, Theory and Applications* (Springer) p. 322

[5] Baig I, Ramesh K and Hariprasad MP 2015 Analysis of stress distribution in dry masonry walls using three fringe photoelasticity *Proc. SPIE 9302, Int. Conf. on Experimental Mechanics 2014* November 15, 2014 Singapore

[6] Pan B, Qian K, Xie H and Asundi A 2009 Two-dimensional digital image correlation for in-plane displacement and strain measurement: a review *Meas. Science & Techn.* **20** (6) 062001

[7] Herbert DM, Gardner DR, Harbottle M, Thomas J and Hughes TG 2011 The development of a new method for testing the lateral load capacity of small-scale masonry walls using a centrifuge and digital image correlation, *Constr. & Build. Mat.* **25** 4465–76.

[8] Vanniamparambil PA, Khan F, Schwartz E, Kontsos A, Bartoli I, Bolhassani M and Hamid A. 2013 Using DIC to measure deformation fields of concrete Masonry test specimens *Proc. of 12th Canadian Masonry Symp.* June 2-5 Vancouver, British Columbia, 333-1173-2-DR.

[9] Ramos R, Furtado A, Eslami S, Alves S, Rodrigues H, Arêde A, Tavares PJ and Moreira PMGP 2015 2D and 3D Digital Image Correlation in Civil Engineering – Measurements in a Masonry Wall *Proc. Engineering* **114** 215-222

[10] Salmanpour AH, Mojsilovića N, Schwartz J 2015 Displacement capacity of contemporary unreinforced masonry walls: An experimental study *Eng Struct* **89** *1-16*

[11] Post D 1979 Photoelasticity *Exp. Mechanics* **19** (5) 176-192

[12] Mueller H 1938 The theory of photoelasticity *J American Ceramic Society* **21** 27–33

[13] Philips JW 2000 TAM 326 – Experimental Stress Analysis-Photoelasticity (Board of Trustees University of Illinois at Urbana-Champaign) p 62

[14] Sutton MA, Yan JH, Tiwari V, Schreier HW and Orteu JJ 2008 The effect of out-of-plane motion on 2D and 3D digital image correlation measurements *Optics and Lasers in Engineering* **46** (10)

[15] Gabrielli E and Colla C 2014 Monitoring of masonry compression tests in the lab via optical correlation without surface preparation, *Key Eng. Mat.* **624** 139 – 146

[16] HuňadyR, Hagara M and Trebuňa F 2014 The Measurement of Standing Wave Patterns by using High-speed Digital Image Correlation *Amer. J Mech. Engineering*, **2** (7) 247-251

[17] Roberts TM, Hughes TG, Dandamudi VR and Bell B 2006 Quasi-static and high cycle fatigue strength of brick masonry *Constr.& Build. Mat* **20** 603-14

[18] Brencich A, Corradi C and Gambarotta L 2008 Eccentrically loaded brickwork: theoretical and experimental results *Eng. Struct.* **30** 3629-43

[19] Adam JM, Brencich A, Hughesc TG and Jefferson T 2010 Micromodelling of eccentrically loaded brickwork: study of masonry wallettes *Eng. Struct.* **32** 1244-51

[20] Brencich A and de Felice G 2009 Brickwork under eccentric compression: experimental results and macroscopic models *Constr.& Build. Mat.* **23** 1935–46
[21] D.M. 14/01/2008 Italian code for construction (NTC 2008) *in Italian language*
[22] www.vishaypg.com
[23] Colla C, Gabrielli E and Carli R 2015 Optical innovative monitoring of masonry testing on site *Proc. of AIPND 2015*, 21-23 Oct. 2015 Milano Italy IDN30 pp 10
[24] VIC-3D 2012 Manual & Testing Guide, www.correlatedsolutions.com

Mini-DIAL system measurements coupled with multivariate data analysis to identify TIC and TIM simulants: preliminary absorption database analysis.

P Gaudio[1], A Malizia[1], M Gelfusa[1], E. Martinelli [2], C. Di Natale[2] , L.A. Poggi[1], C. Bellecci[3].

[1]Department of Industrial Engineering, University of Rome "Tor Vergata", Faculty of Engineering, 00133, Rome, Italy
[2]Department of Electronic Engineering, University of Rome "Tor Vergata", Faculty of Engineering, 00133, Rome, Italy
[3]Crati s.c.r.l. c/o Unical, Rende (CS), Italy.

E-mail: gaudio@ing.uniroma2.it

Abstract. Nowadays Toxic Industrial Components (TICs) and Toxic Industrial Materials (TIMs) are one of the most dangerous and diffuse vehicle of contamination in urban and industrial areas. The academic world together with the industrial and military one are working on innovative solutions to monitor the diffusion in atmosphere of such pollutants. In this phase the most common commercial sensors are based on "point detection" technology but it is clear that such instruments cannot satisfy the needs of the smart cities. The new challenge is developing stand-off systems to continuously monitor the atmosphere. Quantum Electronics and Plasma Physics (QEP) research group has a long experience in laser system development and has built two demonstrators based on DIAL (Differential Absorption of Light) technology could be able to identify chemical agents in atmosphere. In this work the authors will present one of those DIAL system, the miniaturized one, together with the preliminary results of an experimental campaign conducted on TICs and TIMs simulants in cell with aim of use the absorption database for the further atmospheric an analysis using the same DIAL system. The experimental results are analysed with standard multivariate data analysis technique as Principal Component Analysis (PCA) to develop a classification model aimed at identifying organic chemical compound in atmosphere. The preliminary results of absorption coefficients of some chemical compound are shown together pre PCA analysis.

Keywords - Dial; TIC, TIM,

1. Introduction

Nowadays the risk of a release of a chemical release is higher than in the past [1]. The real challenge is getting not only a better knowledge on risks, agents, protection-decontamination [2], detection and identification and investigation techniques, but also the establishment of a doctrine on prevention capabilities (referred to new non-proliferation methodologies), and learning to face non-conventional events and to manage their very consequences. The development of new technologies to detect and identify chemical agents (CA) will be the focus of the paper, the author will approach new methodologies to face this problem [3].

The CAs release may be divided into two main categories: natural events (that provokes accidents having as consequences the CAs dispersion), man-made accidents that are respectively divided unintentional or intentional. [4]

One the most well-known unintentional events was the one occurred in 1976 in Seveso, Italy, where a dense vapor cloud was released from a chemical plant manufacturing pesticides and herbicides. In Europe, the Seveso accident prompted the adoption of legislation aimed at the prevention and control of such accidents. The toxic cloud contained tetrachlorodibenzoparadioxin (TCDD), a by-product of the trichlorophenol synthesis, also known as Seveso dioxin. TCDD has poisonous and carcinogenic properties with an LD_{50} of 0.02 mg/kg. Although no fatalities were reported, soon after its release a large amount of different toxic chemicals were dispersed in the environment and spread on a large area. This resulted in the immediate contamination of some ten square miles of land and vegetation. More than 600 people had to be evacuated from their housings and as many as 2000 were treated for dioxin poisoning [5] Moving to recent years, another chemical, accident related event is the one occurred in Viareggio, Italy, in 2009. The Viareggio derailment of a freight train and subsequent fire occurred on June 29, 2009, in a railway station in Viareggio, (province of Lucca), a city in Central Italy's Tuscany region. Some of the wagons were reported to have been carrying Liquefied Petroleum Gas (LPG). Two of these exploded and caught fire. Seven people were reported to have died in a building collapse.

In order to face the problem of a CA contamination there are several actions that have to be taken, the capability to rapidly detect and identify the contamination is an imperative. This knowledge allows the severity and extent of a hazard to be assessed so that areas that are clean or contaminated can be identified. Furthermore, the information acquired by these systems provides advice to military commanders and first responders, regarding the donning of individual protective equipment (IPE), sampling, handling and analysis procedures as well as medical countermeasures, should the need arise. An ideal detector can be described as one that can detect both CWAs and toxic industrial chemicals (TICs)/toxic industrial materials (TIMs) selectively within an acceptable time; sensitive enough to detect agent concentrations at or below levels which pose a health risk, and not be affected by other factors in the environment. The detector should have a rapid reaction and recovery time whilst being portable, easy to operate and produce data that is easily interpreted. As yet, the 'ideal' detector is not a commercial reality.

The Quantum Electronics and Plasma Physics Research Group[1] (QEP) of Department of Industrial Engineering has a long experience in stand off detection and identification development of apparatus based on laser systems [6-16] and in this paper a new demonstrator developed based on differential light absorption principle will be presented together with a PCA technique for data analysis applied by the our Department of Electronic Engineering. Preliminary result of absorption coefficient database joint with discrimination analysis will be reported and discussed.

2. Material and methods

Optical remote sensing, remote analysis by means of light, consist in a two-steps analysis: the first step consists in a proper excitation of the medium under study by sending into it radiation at optical frequencies, the second step consists in collecting from it a suitable light signal which has the property of being a direct or indirect carrier of the information needed about the interaction. Therefore, we need a suitable light source as a light transmitter. The receiving antenna usually could be a lens/objective or a telescope. It has the task to collect the light on a detector for the conversion of the optical signal in an electrical one in which is much easier to handle [17-29].

[1] www.qepresearch.it

2.1 Experimental demonstrator and its principle of work

The literature present [17-29] affirms that the LIDAR (Light Detection And Ranging) techniques can be used only to provide an alarm in case of chemical release. In case of chemical events, it is essential to deploy also systems able to identify the chemical substances released. For this reason, the QEP research group are developing a mini-Dial system multi-wavelengths based named TELEMACO. The Dial technology usually employs two different wavelengths in order to measure the concentration profiles of an investigated molecule. In the case of organic substance unfortunately the absorption spectra show different overlaps causing interference phenomena that do not allow the recognition using the classical dial methodologies based only couples of laser lines [30]. The Dial multi-wavelengths methodologies, instead could be applied in order to identify the organic compounds using backscattering signals from different wavelengths Consequently, a fingerprint database is required, composed by a collection of absorption coefficients to identify molecules and to minimize the risk of wrong detections or false alarms. Actually data mining techniques have been developed to analyse the database of fingerprints in order to selected, for every organic compound, the minimum number of laser lines essential to identify the gas minimizing false alarm conditions.

Nowadays there is not such a collection of data in the scientific and technical literature. Furthermore the exactly value of absorption coefficients is influenced by the type of light source adopted. The mini TEA CO_2 laser (characteristics reported in table 1) is adopted for mini DIAL system TELEMACO. Preliminarily the mini TEA CO_2 laser is been used for absorption coefficient measurements made with cells filled with the investigated molecules. In this way we have obtained a preliminary results of proper fingerprint database, necessary for future application in atmosphere in order to identify dangerous molecules.

Table 1. TEA CO_2 Laser characteristics.

Transmitter TEA CO_2 laser	
Output power	10^9 W
Beam divergence	0.77 mrad
Spectral range	9:11 μm
Receiver	
Primary ROC	1300 mm
Primary diameter	250 mm
ZnSe focal length	10 mm
Detector type	HgCdTe
Detector Sensitivity D*	$3.38 \cdot 10^{10}$ cmHZ$^{1/2}$/W
Detector size	1 mm^2

There are two principal reason because the CO_2 laser has been chosen: first it is eye safe and, second because it covers a spectral band where there is good absorption for this kind of molecules. The preliminary measurements are carried out in laboratory in order to evaluate the different spectrum of several aggressive chemical compounds. A picture of absorption layout assembled in laboratory for in cell measurements using the same laser of TELEMACO is shown in Figure 1:

Figure 1. Laboratory absorption cell apparatus.

The absorption setup is based on a TEA CO_2 laser tuneable over 60 wavelengths, from 9 to 11 μm, and it has been fully equipped with an electronic micrometer to automatically rotate the diffraction grating and then is able to change the wavelengths.

The absorption coefficient $\alpha(\lambda_i)$ for the i-esima λ_i is calculated by the ratio between the transmitted I_T beam energy and the measured one by the sensor number 2 (Transmitted intensity - I_T) and by the sensor number 1 (Incident intensity - I_0). Indicating with β the reflectance of the beam splitter, the equations 1, 2 and 3 relate the three intensities:

$$I_1 = \beta \cdot I \quad (1)$$

$$I_0 = (1 - \beta) \cdot I \quad (2)$$

$$I_T = I_0 A_f A_g = (1 - \beta) \cdot I \cdot A_f A_g \quad (3)$$

where I is the intensity of the laser pulse Incident (I_1) reflected Intensity and (I_0) transmitted separated by beam splitter. The transmitted pulse is sent on absorption cell given a transmitted intensity (I_T) by cell and A_f and A_g represents respectively the attenuation of the laser beam due to the windows of the cell and the gas inside the cell (that is the CAs). In order to evaluate the gas attenuation it is need evaluate before the absorption due to at layout adopted as attenuation of the bean splitter and wavelength dependences or conversion coefficients of energy monitor. The dependency of the values β and A_f and this can be achieved studying the response of an empty cell (CV) and calculating for this particulate case the ratio I_T/I_1. The solution at this problem in the follow equation is given:

At the first the transmitted intensity of at empty cell is measured:

$$I_T^{CV} = I_0 A_f = (1-\beta)\cdot I\cdot A_f \quad (4)$$

So the ratio measured with the empty cell will be (5):

$$\left(\frac{I_T}{I_1}\right)_{CV} = \frac{(1-\beta)\cdot I\cdot A_f}{\beta\cdot I} = \frac{A_f}{\beta} \quad (5)$$

Then the value is computed for the CAs inside the cell (6):

$$\left(\frac{I_T}{I_1}\right)_{CP} = \frac{(1-\beta)\cdot I\cdot A_f\cdot A_g}{\beta\cdot I} = \frac{A_f}{\beta}\cdot A_g \quad (6)$$

So it is possible evaluate the attenuation A_g (7):

$$\frac{\left(\frac{I_T}{I_1}\right)_{CP}}{\left(\frac{I_T}{I_1}\right)_{CV}} = A_g \quad (7)$$

The absorption value, $\alpha(i)$, is related to the attenuation value by eq. (8):

$$A_g(i) = exp(-\alpha(i)\cdot p\cdot l) \quad (8)$$

Where p is the partial pressure of the gas, in [atm], and l is the length path of the laser beam into the cell, in [cm]. So it is possible to obtain the absorption value, $\alpha(i)$ (9):

$$\alpha(i) = \frac{1}{p\cdot l}\ln\frac{1}{A_g} \quad (9)$$

and the relative error (10):

$$\frac{\Delta\alpha(\nu)}{\alpha(\nu)} = \frac{\Delta P}{P} + \frac{\Delta l}{l} + \frac{I_1}{I_T} * \frac{1}{\ln\left(\frac{I_T}{I_1}\right)} * \Delta\left(\frac{I_T}{I_1}\right) \quad (10)$$

2.2 Multivariate data analysis method

Exploiting the information content of the multi-wavelength approach of the LIDAR, each single measurement is characterized by a pattern of absorption coefficients that represent a sort of fingerprint of the volatile compounds composing the air. To analyze this multidimensional data we need to process the experimental measurements by means of multivariate data analysis techniques.

Multivariate analysis consists of a collection of algorithms that can be used when measurements are characterized by a collection of descriptors. Historically, the great part of applications of multivariate techniques have regarded biological and behavioral sciences. Nevertheless, interest in multivariate methods has now spread to different fields of investigation. In this work we have analyzed the experimental results with two different data analysis techniques. The first one is an explorative and unsupervised technique called Principal Component Analysis (PCA).

PCA is a decomposition technique unsupervised used in multivariate data to obtain uncorrelated components. With PCA is maximized the variance of a linear combination of different features and allows to reduce the number of variables and to detect the presence of structure in the relationships between them.
Using PCA is possible to represent a dataset in a graphical representation carrying most of the data variance [32]. In the new representation space, the axes are obtained from a linear combination of the original axes and they are requested to be uncorrelated and orthogonal. In the direction of the largest variance of data is chosen the firs axis, called first "Principal Component". The second Principal Component orthogonal direction to the first PC with largest variance, and so on [32]. The second techniques is k-NN k- Nearest Neighbors one the golden standard classifiers [33]. k-NN is a simple algorithm that classifies new cases based on a similarity measure of k nearest neighbours selected among a pool of training examples.

3. Results

3.1 Experimental results and discussion

The absorption laboratory layout based on the same CO_2 laser source that will be used in TELEMACO DIAL systems is used to get the absorption spectra (see some of them in Figure 2, 3 and 4) that are principal characteristic of each CA.

The following figures 2,3,4 are three examples of the absorption spectra for toluene, trimethylbenzene and n-heptane taken placing few ppm of substance in the cell in two different conditions:

- Pure substances: condition obtained adding the substance in a cell after that a high vacuum has been reached (in the order of 0,1 bar)
- Substance in the air: air is added in the cell starting from the above mentioned set-up till the pressure of 1 atm is reached inside the cell.

These examples shows how clear are the absorption spectra for each substance both in "pure condition" and in "real scenario conditions". The differences between the absorption value of pure gas and an experimental condition closer to reality (air + substances) can be explained with the increase of internal pressure in this second condition,

The small discrepancy between the absorption values measured in two different conditions justifies the use of CAs + air to determinate the absorption value in order to realise a database of fingerprints. It is the starting point to get a fully database of absorptions spectra of the chemical aggressive agents indispensable for applied the dial techniques in order to identify chemical gas in atmosphere.

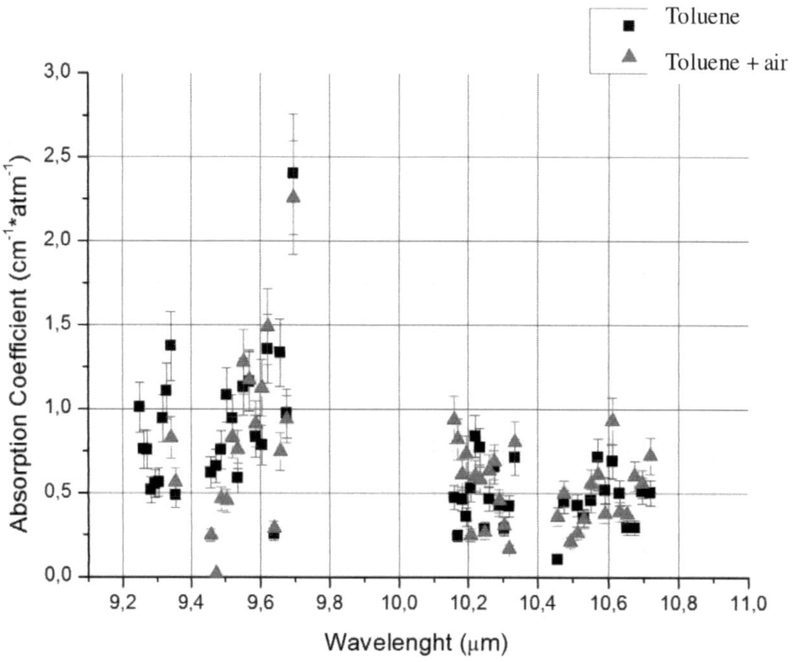

Figure 2. Absorption coefficients of toluene

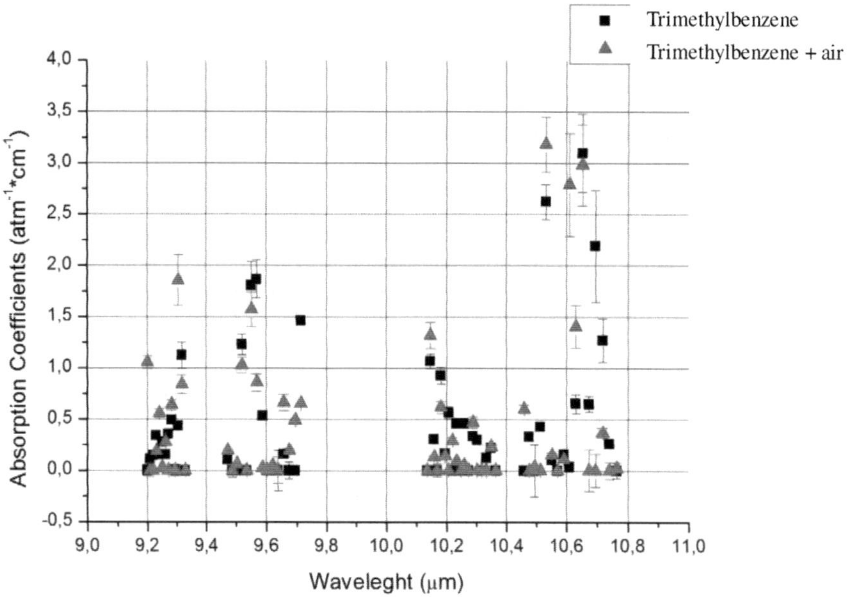

Figure 3. Absorption coefficients of trimethylbenzene

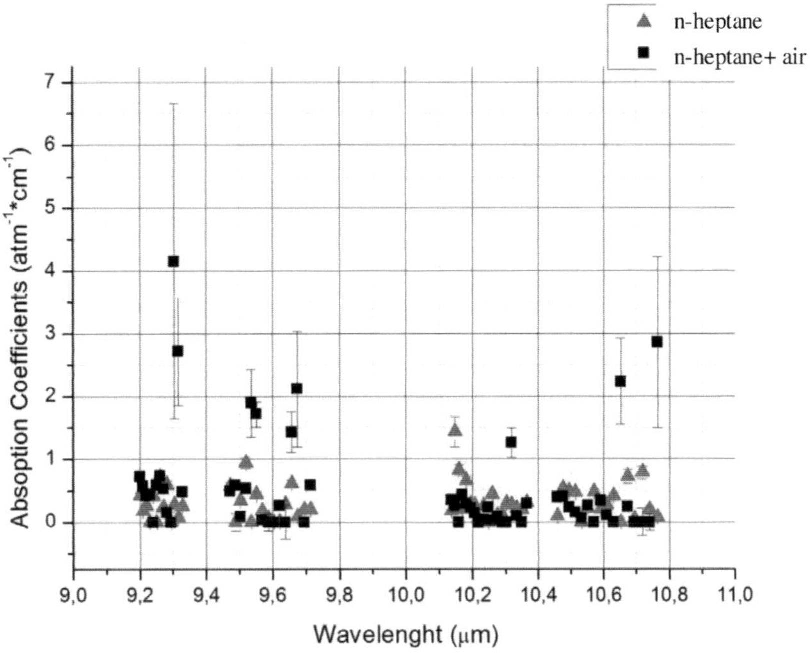

Figure 4. Absorption coefficients of n-heptane

Observing the figures 2,3,4 is possible assume that :

- The CO_2 laser source is able to produce the absorption spectra that are clearly different and constitute the finger-prints of the TIM/TIC used for the experiments;
- The absorption of the pure substance is almost always higher than the one of substance+air it means that the presence of air provoke a decrease in the absorption coefficient. The behaviour of the spectra is almost the same.

3.2 Discussion of data elaborated with multivariate analysis

Figure 5 shows the scores plot of the first two principal component of the PCA model built with the dataset related to several replicas of 5 different compounds in air:

- Allylic alcohol
- Benzene
- Chlorobenzene
- Methanol
- Piperidine

It is interesting to observe that the different volatile compounds are clearly separated confirming the potentiality of the proposed approach as method to distinguish the different compounds. These discrimination capabilities is further confirmed by the k-NN classification model that obtains a perfect discrimination score for all the 5 classes with leave 5 out cross validation procedure[33].

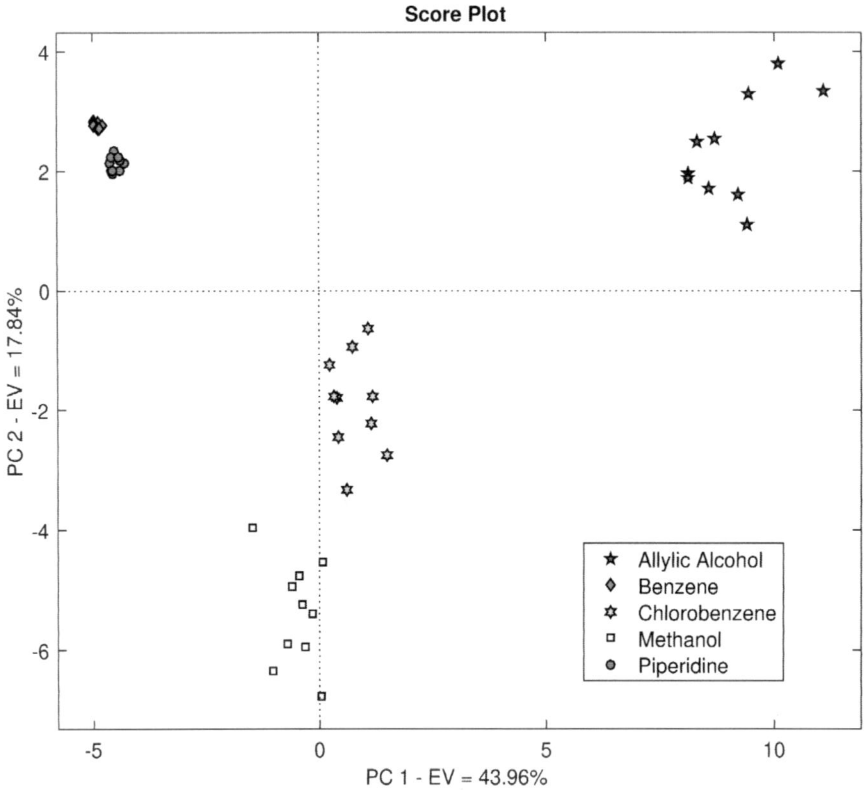

Figure 5. Scores plot of the PCA model built with dataset related to the five measured compounds

4. Conclusions

During the last fifteen years, the authors have demonstrated that laser-based optical techniques can enable a continuous, and quasi real-time, monitoring of particulate and of the CAs, in air for medium-long distances. In particular, the capability of LIDAR systems to detect CAs for the early warning phase and that of the DIAL systems could be used to identify at short-medium range some CAs, so the usefulness of those systems to support (once that the test in field will be done) decision making. The risks of chemical contaminations can be relevant also in healthcare settings and in a number of industrial plants; therefore, it is necessary to guarantee, in all these environments, a continuum monitoring to reduce the probabilities and the magnitude of harmful events. Once that the absorption spectra are obtained it has been necessary find a way to identify the substances starting from the data matrix. The method applied based on PCA and the K-NN classification method allow the identification of all the substances, it means that once prepared a database with all the absorption spectra of the substances it will be possible pass to the identification phase through the mentioned approach. The authors have demonstrated that the use of optical techniques could be a viable solution with a an experiment aimed at identifying five different volatile compounds in air showing promising results. The challenges for the future are: to optimize the systems reducing the dimensions and the costs, while improving at same time the performances in terms of reduction of false alarms, to increase the robustness, the acquisition speed and the reliability during the acquisition and elaboration data phases.

Acknowledgements

A special acknowledgements to the students Roberto Palmieri and Luigi Prestianni for his support under the experimental campaigns.

References

[1] Malizia A. et all, 2016. Disaster management in case of CBRNe events: an innovative methodology to improve the safety knowledge of advisors and first responders. *Defense and Security Analysis*. **32** (1), 79-90.
[2] Sassolini A. et all, 2014. Evaluation of the effectiveness of titanium dioxide (TiO2) Self-cleaning coating for increased protection against CBRN incidents in critical infrastructures. *Defence S and T Technical Bulletin*. **7** (1), 9-17
[3] Lungaroni M. et all., 2016. New analysis methods to push the boundaries of diagnostic techniques in the environmental sciences. *Journal of Instrumentation*. **11** (4), 16.
[4] Iannotti A. et all, 2016. Weapons of mass destruction: A review of its use in history to perpetrate chemical offenses. *Defence S and T Technical Bulletin*. **9** (1), 39-52.
[5] http://ec.europa.eu/environment/seveso/
[6] Bellecci C, Caputi G, De Donato F, Gaudio P and Valentini M 1995 CO2 Dial for monitoring atmospheric pollutants at the University of Calabria, *11° Nuovo Cimento* **18** 5 463-472
[7] Bellecci C, Francucci M, Gaudio P, Gelfusa M, Martellucci S, Richetta M and Lo Feudo T 2007 *Appl. Phys*. **B 87** 373.
[8] Bellecci C, Gaudio P, Gelfusa M, Lo Feudo T, Malizia A, Richetta M, Ventura P, 2009. Raman water vapour concentration measurements for reduction of false alarms in forest fire detection 2009, *Proceedings of SPIE - The International Society for Optical Engineering* vol. **7479**.
[9] Bellecci C, Gaudio P, Gelfusa M, Lo Feudo T, Murari A, Richetta M and De Leo L 2010 In-cell measurements of smoke backscattering coefflcients using a CO2 laser system for application to Lidar-dial forest fire detection *Optical Engineering* **49** (12) 124302

[10] Bellecci C, Gaudio P, Gelfusa M, Malizia A, Richetta M, Serafini C and Ventura P 2010 Planetary boundary layer (PBL) monitoring by means of two laser radar systems: experimental results and comparison *SPIE Conference* vol. **7832**
[11] Gaudio P, Gelfusa M, Lupelli I, Malizia A, Moretti A, Richetta M, Serafini C and Bellecci C 2011 First open field measurements with a portable CO2 lidar/dial system for early forest fires detection, *SPIE Conference* vol. **8182**
[12] Gaudio P, Gelfusa M, Malizia A, Richetta M, Serafini C, Ventura P, Bellecci C, De Leo L, Lo Feudo T and Murari A 2013 New frontiers of Forest Fire Protection: A portable Laser System (FtED), *WSEAS TRANSACTIONS ON ENVIRONMENT AND DEVELOPMENT*, vol. **9** 195-205
[13] Gaudio P, Gelfusa M, Malizia A, Richetta M, Antonucci A, Ventura P, Murari A and Vega J 2013 Design and development of a compact Lidar/Dial system for aerial surveillance of urban areas, *SPIE Conference* vol. **8894**
[14] Gelfusa M, Malizia A, Parracino S, Richetta M, Gaudio P, Murari A and Vega J, 2014 Automatic localization of backscattering events due to particulate in urban areas *SPIE Conference*, Amsterdam 24-26 September vol. **9244**.

[15] Gelfusa M, Gaudio P, Malizia A, Murari A, Vega J, Richetta M, Gonzalez S 2014 UMEL: A new regression tool to identify measurement peaks in LIDAR/DIAL systems for environmental physics applications *Review of Scientific Instruments* Volume **85**, Issue 6, Article number **063112**

[16] Gelfusa M, Malizia A, Parracino S, Richetta M, Bellecci C, Avolio E, De Leo L, Perrimezzi C and Gaudio P 2014 Detection of pollutant sources in the atmosphere with Lidar/Dial techniques: Results of an experimental campaign in the south of Italy (Naples: Fotonica Conference) (in Italy)

[17] Measures R 1984 *Laser Remote Sensing – Fundamentals and Applications* (Florida: Krieger) (in USA)

[18] Weitkamp C 2005 *LIDAR: Range-Resolved Optical Remote Sensing of the Atmosphere* (Germany: Springer)

[19] Kovalev V 2004 *Elastic LIDAR, Theory, Practice and Analysis Methods* (Berlin: Wiley) (in German)

[20] Schotland R 1974 Errors in the lidar measurements of atmospheric gases by differential absorption J. *Appl. Meteor.* **13** 71-77

[21] Andreucci F and Arbolino M 1993 A study on forest fire automatic detection system, 2 – smoke plume detection performance *Il Nuovo Cimento* **16** 1-51

[22] Utkin A, Fernandes A, Costa L, Vilar R and Simoes F 2002 Detection of small forest fire by lidar *Appl. Phys B Lasers Opt* **74** 1 77-83.

[23] Utkin A, Fernandes A, Costa L, Vilar R and Lavrov A 2002 Forest fire detection by means of lidar, Forest fire research and wildland safety *Proceedings of the IV International Conference on Forest Fire Research*, Rotterdam: Millpress, 58

[24] Byer L 1975 Remote air pollution measurements *Optical and Quantum Electronics* Vol.**7** 147-177

[25] Kildar H and Byer L 1971 Comparison of laser methods for the remote detection of atmospheric pollutants *proceeding of IEEE* vol. **59** 1644-1663

[26] He T Y, Stanic S and Gao F 2012 Tracking of urban aerosols using combined LIDAR-based remote sensing and ground-based measurements (University of Nova Gorica) (in Slovenia)

[27] Del Guasta M 2002 Daily Cycles in urban aerosol observed in Florence by means of an automatic 532-1064 nm LIDAR (IROE CNR, Firenze) (in Italy)

[28] Bellecci C, D'Aquila G, Gabriele S, Gaudio P, Martellucci S, Ribecco G and Richetta M 2003 Comparison of columnar water vapor measurements using the CO2 DIAL method and gps data analysis, *EUROPTO European Symposium on Remote Sensing*, 8 September, Barcellona, Spain, vol. **5232**, 521-527

[29] Menyuk N, Killinger D K and De Feo E 1982 Laser remote sensing of hydrazine, MMH and UDMH using a differential absorption CO2 Lidar *Appl. Optics* **21** 12-2275

[30] Measure R. M., Laser Remote Sensing, Krieger, 1984.

[31] Bellecci C. Gaudio, P., Gelfusa, M., Martellucci, S., Richetta, M., Ventura, P., Antonucci, A., Pasquino, F., Ricci, V., Sassolini, A., 2008. Database for chemical weapons detection: First results. *Proceddings of SPIE*, 7116.

[32] Martens, H, & Naes, T. (1989). Multivariate Calibration. London: Wiley.

[33] Bishop CM . *Pattern Recognition and Machine Learning* .Heidelberg, Germany: Springer ; 2006

Water Spray Flow Characteristics Under Synthetic Jet Driven By a Piezoelectric Actuator

L. Marchitto[1], G. Valentino[1], M. Chiatto[2], L. de Luca[2]

[1] Istituto Motori, CNR - NAPOLI – ITALY,
[2] Dept. of Industrial Engineering, University of Naples "Federico II" - NAPOLI - ITALY

E-mail: l.marchitto@im.cnr.it

Abstract. Particle Image Velocimetry (PIV) and Phase Doppler Anemometry (PDA) have been applied to investigate the droplets size and velocity distribution of a water spray, under the control of a piezo-element driven synthetic jet (SJ).
Tests were carried out under atmospheric conditions within a chamber test rig equipped with optical accesses at two injection pressures, namely 5 and 10 MPa, exploring the variation of the main spray parameters caused by the synthetic jet perturbations.
The SJ orifice has been placed at 45° with respect to the water spray axis; the nozzle body has been moved on its own axis and three different nozzle quotes were tested. PIV measurements have been averaged on 300 trials whereas about 10^5 samples have been acquired for the PDA tests. For each operative condition, the influence region of the SJ device on the spray has been computed through a T-Test algorithm.
The synthetic jet locally interacts with the spray, energizing the region downstream the impact. The effect of the actuator decreases at higher injection pressures and moving the impact region upwards. Droplets coalescence can be detected along the synthetic jet axis, while no significant variations are observed along a direction orthogonal to it.

1. Introduction

Spray-based technology has been undergoing continuous evolution in the last decades, covering a wide spread of applications such as internal combustion engines, turbines, fire control systems, as well as agriculture machines and several manufacturing processes. The technological improvement has been mainly focused on pressure control, size and nozzle-hole number, tip penetration length and cone angle. At the same time, the opportunity of directly managing spray formation and its characteristics has gained much interest among active flow control methods using different types of actuators [1]. A first classification of spray control mechanisms was made by Photos and Longmire [2] depending on the way of interacting with a spray: direct, if a direct impact between the control mechanism and droplets occurs; indirect, if the interaction is produced by controlling the space and the size of fluid structures. Among them, synthetic jet actuators seem to be a promising technology for spray direct control, showing the potential to affect their global behavior as well as their characteristics at local level.

As depicted in Figure 1, a synthetic jet (SJ) is an electromechanical device consisting of a relatively small cavity, which is sealed from one side by an elastic vibrating diaphragm, and from the other one, is linked to the external environment through a slot or an orifice. The diaphragm (membrane, wall) oscillation, generally driven by a piezo-ceramic element, produces periodic cavity volume changes, with corresponding pressure variations, that cause subsequent alternation of ejection and suction phases of fluid across the orifice, [3,4,5]. During the expulsion phase a vortex ring forms near the orifice exit which, under favorable operating conditions [6], convects downstream by its self-induced velocity towards the far field. Few cycles are required for the formation of a train of vortex rings that interact with each other and break up due to the viscous dissipation eventually "synthesizing" a turbulent jet always directed downstream, [7]. A major characteristic of this jet lies in a zero-net mass-flux during an operating cycle, but in a non-zero momentum flux. Furthermore, it does not require any continuous supply of fluid for its generation, because it is synthesized directly from the surrounding fluid.

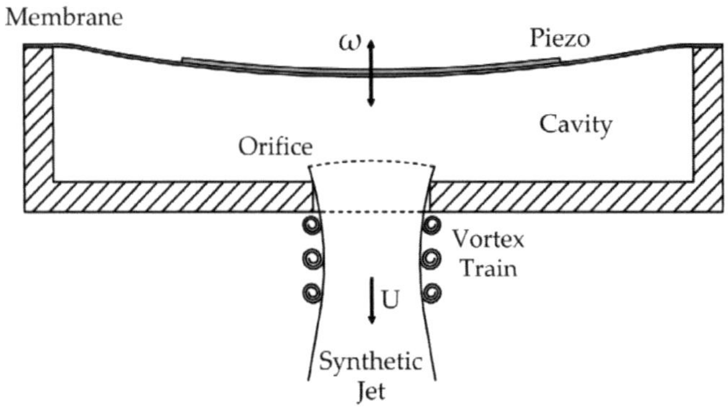

Figure 1. Sketch of the synthetic jet device.

The effect of a SJ actuator on an air atomized spray was investigated by Pavlova et al. [8], quantified through a non-dimensional parameter: the momentum coefficient, defined as:

$$C_\mu = \frac{A_j \overline{U}_j^2}{A_s \overline{U}_{so}^2} \tag{1}$$

where A_j and A_s are the orifice area of the synthetic jet and the spray, respectively, whereas $\overline{U}_{so}$ is the average air exit velocity at the spray orifice, and $\overline{U}_j$ is the synthetic jet average orifice velocity. The latter is known as:

$$U_j = \frac{L}{T} \tag{2}$$

in which T is the actuator period and L is the stroke length, defined as the integral of the velocity at the orifice exit (spatially averaged), U_j, over the ejection phase of the cycle only:

$$L = \int_0^{T/2} U_j(t)dt \tag{3}$$

This paper aims to investigate the effect of a synthetic jet actuator on a continuous water spray. The experimental apparatus, the methodologies applied and the characteristics of the actuator will be presented in Section 2. Section 3 will be dedicated to the definition of a new momentum coefficient, which takes into account the densities of the operating fluids, and it is followed by the results in terms of spray velocity and droplets distribution fields. Finally, conclusions will be reported in Section 4.

2. Experimental Apparatus and Procedure

The spray behavior has been characterized through optical diagnostic techniques. Considering the low droplets velocity and size, a statistical analysis has been applied to establish the spray regions that have been effectively influenced by the SJ device.

Particle Image Velocimetry (PIV) and Phase Doppler Anemometry (PDA) techniques have allowed to measure spray velocity and droplets mean diameter, respectively. Figure 2 shows the PIV experimental apparatus and the water injection system. The latter is composed of a high pressure pump used to supply water to a one-hole misting nozzle (300 μm diameter), which injects the working fluid in a 4500 cm^3 quiescent optically accessible tank at ambient temperature and pressure. The injector works in continuous mode and its working regime is managed by a rubber ball connected to a spring; when the liquid pressure exceeds the elastic spring force, the ball moves allowing the liquid passage. For this reason the spray can be considered continuous and stationary.

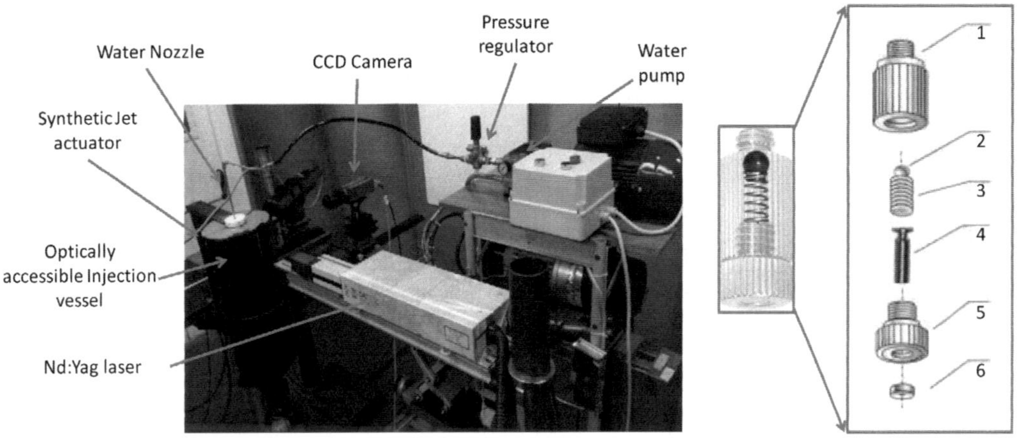

Figure 2. PIV and injection system experimental apparatus (left) - Hole misting nozzle sketch (right).

A twin head pulsed Nd:YAG laser (532 nm wavelength), has been synchronized with a CCD camera in order to perform PIV measurements. An appropriate optical path converts the laser beam in a sheet (0.2 mm of thickness and 60 mm of height) crossing the spray axis, coincident with the vessel one. The light scattered by the spray has been collected by the CCD camera, mounted perpendicularly to it. Images (2048 x 2048 pixel of resolution and 7.4 x 7.4 μm of pixel size) have been acquired in double frame mode in order to record a pair of them for each trigger signal, with a frequency of 1 Hz, for a total of 300 couples of images. An adaptive correlation process has been applied to track the droplets clusters and to estimate their speed. This method uses an initial interrogation area (IA) having a size N times the final IA. Each intermediary result is used as information for the next IA of smaller size, until the smallest IA size is reached. A 32×32 pixels final IA has been reached starting by a size of 128×128 pixels. A local validation procedure has been applied in order to minimize false sample vectors and a 50 % overlap of interrogation areas has been set to recover the loss of vector field resolution. Furthermore, the analysis has been completed by the Peak Height Validation criterion and the Local Neighborhood Validation method.

PDA tests, whose apparatus is shown in Fig. 3, have been performed in the same tank used for PIV experiments. The set-up consists in an argon-ion laser, with operative wavelengths of 514.5 and 488 nm, a transmitting probe with a focal length of 310 mm, beam separation of 65 mm, and receiving optics placed with an angle of 30° with respect to the axis of the transmitting probe in order to maximize the signal-to-noise ratio.

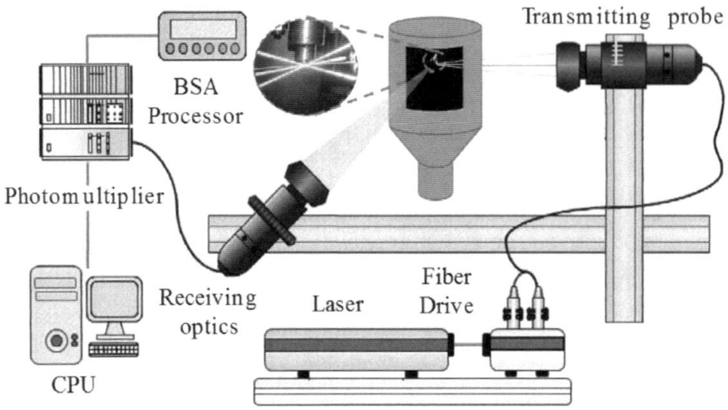

Figure 3. PDA experimental apparatus.

The probes have been mounted on a tridimensional translation slide, allowing the acquisition at different locations inside the spray with micrometric accuracy. An acquisition time of 300 s and a number of samples equal to 100000 were required to collect an adequate data sample for statistical analysis. All the experiments have been conducted in the atmospheric chamber test rig with a drain at the bottom driven by a vacuum pump to remove the moisture. Further details about PIV and PDA equipment can be found in [9].Tests have been carried out under quiescent conditions at two injection pressures, namely 5 and 10 MPa, exploring the effect produced by the synthetic jet perturbations on the droplets velocity and size. The inclination of the synthetic jet axis has been equal to 45° with respect to the nozzle axis direction, as shown in Figure 4. The relative quote of synthetic jet with respect to the water nozzle has been varied from a maximum (Δh = 0 mm) corresponding to 11.6 mm distance between the nozzle and actuator orifices up to 1.6 mm (Δh = 10 mm) with a step of 5 mm.

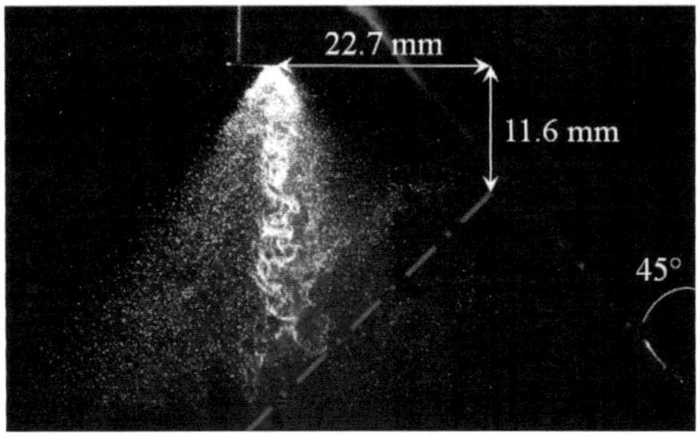

Figure 4. Relative position between the nozzle and the synthetic jet actuator, (Δh=0).

PIV Measurements have been carried out at all set positions, for a total of 12 tests (summarized in Table 1), whereas PDA tests at the intermediate one. More in detail, since PDA allows only punctual measurements, two axis, one coincident with that of the synthetic jet actuator and one perpendicular to it, have been chosen to acquire values.

Table 1. PIV operative conditions.

	Injection Pressure, P_i	Nozzle Height, Δh
Case 1	5 MPa	0 mm
Case 2	10 MPa	0 mm
Case 3	5 MPa	5 mm
Case 4	10 MPa	5 mm
Case 5	5 MPa	10 mm
Case 6	10 MPa	10 mm

The actuator design is based on the work of de Luca et al. [**Errore. Il segnalibro non è definito.**], whose characteristics are reported in Table 2. It is a modular structure device (Figure 5), allowing independent variations of cavity diameter and height, orifice diameter, and piezo-electric diaphragm. Its features and actuation frequency (f=2404 Hz for all experimental tests) have been chosen to maximize the exit jet velocity, and so the momentum coefficient, which is the main parameter for spray interaction.

Table 2. Synthetic jet actuator characteristics.

Cavity Diameter, D	35 mm
Cavity Height, H	3 mm
Orifice length, l	2 mm
Orifice length, d	2 mm
Supply Voltage, V	70 V
Actuation frequency, f	2404 Hz

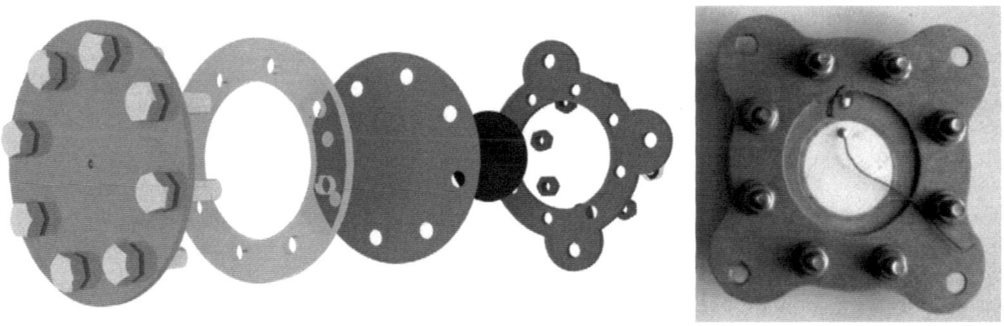

Figure 5. brass synthetic jet actuator (left) - exploded 3D view (right).

Considering the low speed range of water sprays and the statistic nature of PIV measurements, a T-test, implemented in *MATLAB*, has been performed to discriminate between the stochastic spray variability and the perturbation produced by the synthetic jet [10]. For each operative condition two

vector maps, corresponding to active synthetic jet and free water spray cases, have been estimated as the average on 300 trials. A point-by-point comparison of the maps has been performed through the T-test code on the single interrogation areas.
The T-test is based on the assumptions that each sample population X_i follows a normal distribution with mean value μ and variance σ^2 and the two sample populations should have the same variance.
First, a Normal Probability Plot has been built in order to estimate if each set of experimental data follows a normal distribution. The experimental data were sorted in ascending order and for each sorted value the Cumulative Distribution Function (CDF) was evaluated through the following estimator:

$$\hat{F}(x_i) = \frac{i}{n+1} \tag{4}$$

Where x_i is the *i-th* sorted value of sample X and n is the sample dimension. The Cartesian product $(x_i, \hat{F}(x_i))$ has been compared to the linearized analytical CDF distribution. The method of ordinary least squares has been used to approximate the experimental distribution to the analytical one. All the distributions whose correlation coefficient between the experimental values and the least squares line is over than 0.90 have been considered normal. Once the normal distribution of samples has been confirmed, a F-test has been performed for verifying the assumption that the two sample populations have the same variance. The well-known Fisher-Snedecor distribution relation has been used as test statistic:

$$\frac{S_1^2}{S_2^2}\frac{\sigma_2^2}{\sigma_1^2} = Z_{n-1,m-1} \tag{5}$$

where S_1^2 and S_2^2 represent the unbiased sample variance of the aleatory variables X_1 and X_2 respectively, n and m are the sample dimensions. The unbiased sample variance can be written as:

$$S^2 = \frac{\sum_{i=1}^{n}(X_{1i} - \bar{X}_1)^2 + \sum_{i=1}^{m}(X_{2i} - \bar{X}_2)^2}{n + m - 2} \tag{6}$$

where X_{1i} and X_{2i} are respectively the i-th aleatory variables of the sampling of the sample populations X_1 and X_2 which determinate the mean sample aleatory variables $\bar{X}_1$ and $\bar{X}_2$.
The null hypothesis of equality of variances is considered true if the actual value is below 0.05. Once all the assumptions are verified the T-test can be performed through the T-distribution, used as test statistic:

$$\frac{(\bar{X}_1 - \bar{X}_2) - (\mu_1 - \mu_2)}{S\sqrt{\frac{1}{n} + \frac{1}{m}}} = T_{n+m-2} \tag{7}$$

The same threshold as for the F-test has been set for the null hypothesis of equality of means. Hence, if the null hypothesis is not satisfied the means difference can be attributed to the synthetic jet effect. Repeating this analysis for a large number of points in the spray domain, it has been possible to distinguish the region in which the synthetic jet is effective from another where the effect is not relevant, (Figure 6).

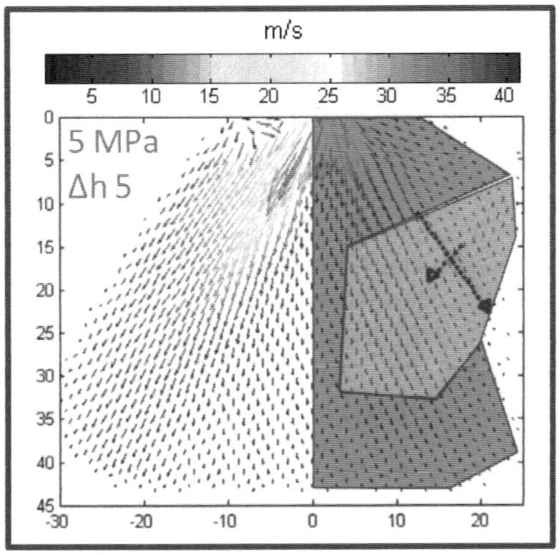

Figure 6. Influence region (blue) of synthetic jet actuator on the averaged velocity vector field at 5 MPa injection pressure, Δh 5 mm.

3. Results and Discussion

3.1. Results and Discussion

The synthetic jet actuator has been designed in order to work in direct mode. For this reason it is important that the air moved by the actuator is strong enough to modify the particles trajectory by means of a direct impact with them. The momentum coefficient C_μ, defined in eq. (1), is the ratio of the synthetic jet momentum to the spray momentum due to the air part only.

In this paper a generalization of this definition is introduced in order to take into account the differences in density between water, used for spray, and air, working fluid of the SJ device. The re-defined momentum coefficient C_μ^* results to be:

$$C_\mu^* = \frac{\dot{m}_j \overline{U}_j}{(\dot{m}_a + \dot{m}_l)\overline{U}_{so}} \tag{8}$$

Where $\dot{m}_j$ is the synthetic jet mass flow rate, $\dot{m}_a$ is the mass flow rate of the air spray component and $\dot{m}_l$ is the mass flow rate of the liquid spray component. Note that if the spray is almost completely composed by air, such as in an air atomized one, the expression can be simplified, recovering the eq. (1). For the present activity C_μ^* have been estimated equal to 4.46 for an injection pressure of 5 MPa and 2.07 for 10 MPa case.

3.2. Spray velocity field

PIV measurements results are reported in Figs. 7-18. In particular, odd figures contain spray velocity fields: the first picture represents the statistical analysis result, whereas the others show a comparison between the zoomed right side of the domain, without and with actuation. The blue region is the influence area of the actuator on the velocity magnitude, while the red one represents the part in which the jet effect is negligible. The arrows, instead, represent the directions along which the velocity profiles have been extracted: the first one coincident with the synthetic jet longitudinal axis, the latter perpendicular to the former. These results are plotted in the even figures.

The jet produced by the device is clearly visible in zoomed pictures due to the presence of water droplets in the injection chamber which, detached from the spray, have been dragged by the air motion and can be considered as tracers. It is interesting to notice a difference in the magnitude of the synthetic jet velocity between the 5 MPa and 10 MPa cases; this can be explained considering the higher atomization rate and the greater amount of water particles at 10 MPa near the synthetic jet orifice caused by the increased flow rate. This results in a higher amount of droplets in the region immediately out of the synthetic jet orifice, where the air speed is generally higher, improving measurements accuracy.

When injection pressure rises up (Figs. 7 and 9), it is possible to note an overall increase in velocity magnitude for both cases, with and without actuation. More in detail, considering the velocity fields, the region of influence (blue one) becomes smaller, according with the two calculated C_μ^* coefficients. As expected at 10 MPa the droplets have higher momentum flux than the case at 5 MPa and so altering their motion becomes more difficult. On the other hand, analysing the velocity profiles, at 5 MPa (Fig. 8) the actuator produces a constant droplets speed increase of about 2 m/s along the longitudinal axis.

At 10 MPa (Fig. 10), instead, the velocity enhancement lowers toward the water spray core, confirming the reduction of the influence. Looking at the transversal axis, for both investigated pressures a significant speed gain has been found. This outcome is characterized by a local maximum located on the synthetic jet axis, while the speed difference between the controlled and uncontrolled cases is almost negligible at about 5 mm distance from the center (borderline of the no-influence region).

The velocity vector fields at 5 and 10 MPa for Δh=5 mm are reported in Figures 11 and 14. Moving downward the injector, the air flow impacts against the spray in a region closer to the nozzle. As expected, the higher water momentum flux induces a reduction of the air flow-water spray interaction, both in terms of interaction region area and of velocity difference. Furthermore, note that at 5 MPa injection pressure the effect of the synthetic jet along its longitudinal axis resulted in a droplet speed increase of about 1 m/s (Figure 12, left).

At 10 MPa injection pressure, such an increase is almost negligible, exhibiting only a slight difference just on the spray edge (between 1 to 2 mm on the spray edge) (Figure 14, left). Looking at the cross axis profiles, as for the Δh=0 mm case, a relatively small velocity increase is detected on the jet axis for injection pressure of 5 MPa (Figure 12, right), while the speed difference is almost negligible at 10 MPa (Figure 14, right).

The difference in the water droplets velocity, between the 5 and 10 MPa configurations, results in a different way in which the jet approaches to the spray. Due to the higher interaction at lower injection pressure, the synthetic jet seems to merge with the spray even upstream the contact point and the result is a slight impact with a totally energization; this allows a global velocity raise and an increase in spray diffusion. On the contrary, at higher injection pressure the impact is sharper because the jet is not able to influence the motion of the droplets, which follow their standard trajectory; this produces a velocity increase limited to the region near the impact point only.

The velocity vector fields at 5 and 10 MPa for Δh=5 mm are reported in Figures 15 and 18. Moving the injector to the lowest position results in a region of air-water impact immediately out of the nozzle hole. On the other hand the distance between synthetic jet orifice and spray is maximum, lowering the air speed in the region of impact. At 5 MPa the air flow-water spray interaction appears to be similar to the case 3, both in terms of interaction region area and of velocity difference. Again, the effect of the synthetic jet along its longitudinal axis resulted in a droplet speed increase of about 1 m/s (Figure 16, left). By increasing the injection pressure, the momentum flux further increases and the effect of the synthetic jet is almost negligible as shown in figure 18. This behavior could be expected considering that the region impact is immediately out of the nozzle hole, where the spray is compact and break up is still occurring. Moreover the higher distance between the synthetic jet actuator and water spray lowers the air velocity in the impact region and consequently its influence.

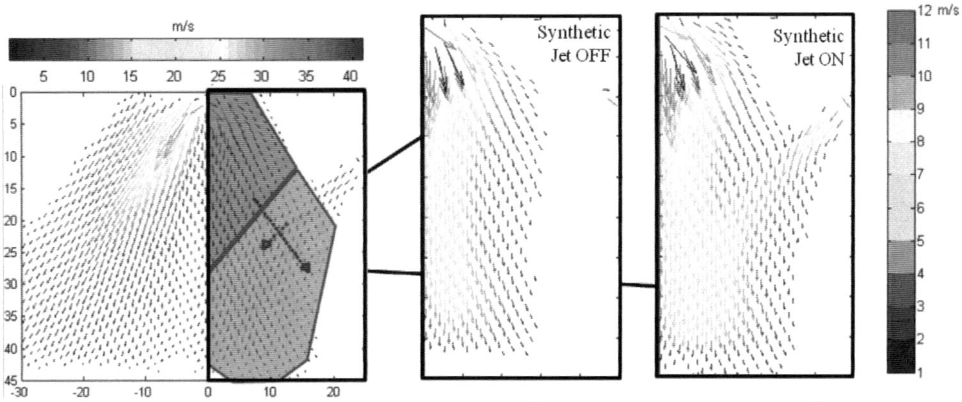

Figure 7. Case 1: Spray velocity field, $(P_{inj} = 5\ MPa, \Delta h = 0\ mm)$.

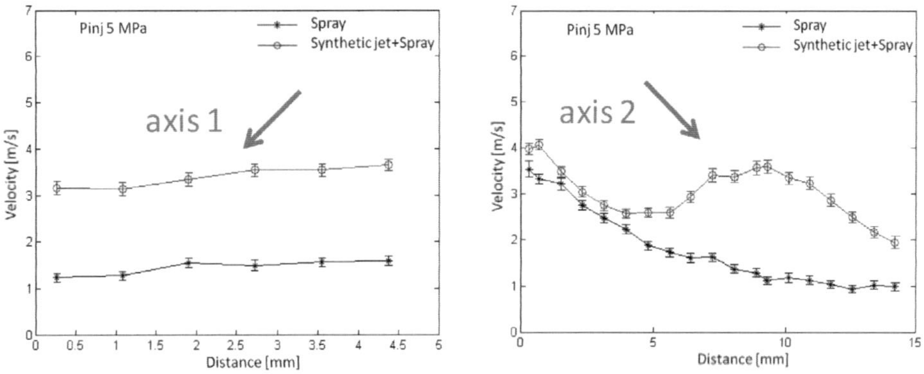

Figure 8. Case 1: Profile plots: longitudinal axis (left) – transversal axis (right), $(P_{inj} = 5\ MPa, \Delta h = 0\ mm)$.

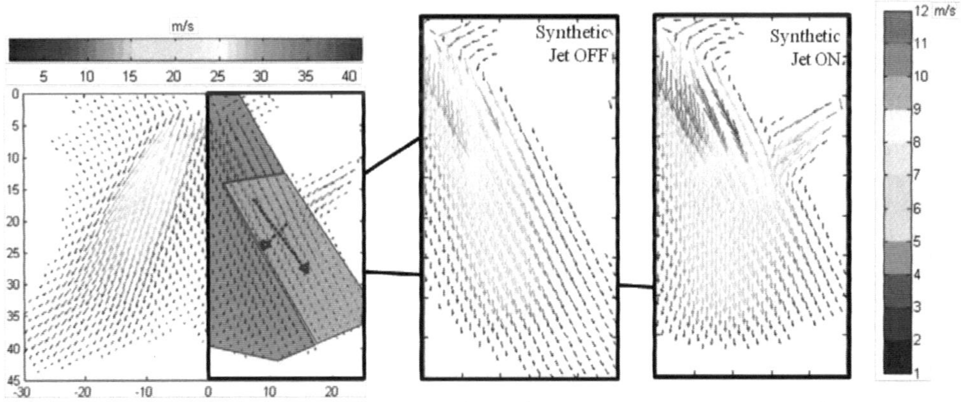

Figure 9. Case 2: Spray velocity field, $(P_{inj} = 10\ MPa, \Delta h = 0\ mm)$.

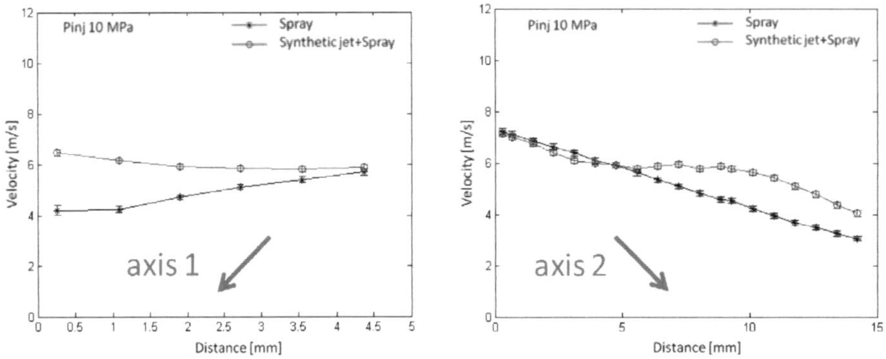

Figure 10. Case 2: Profile plots: longitudinal axis (left) – transversal axis (right), $(P_{inj} = 10\ MPa, \Delta h = 0\ mm)$.

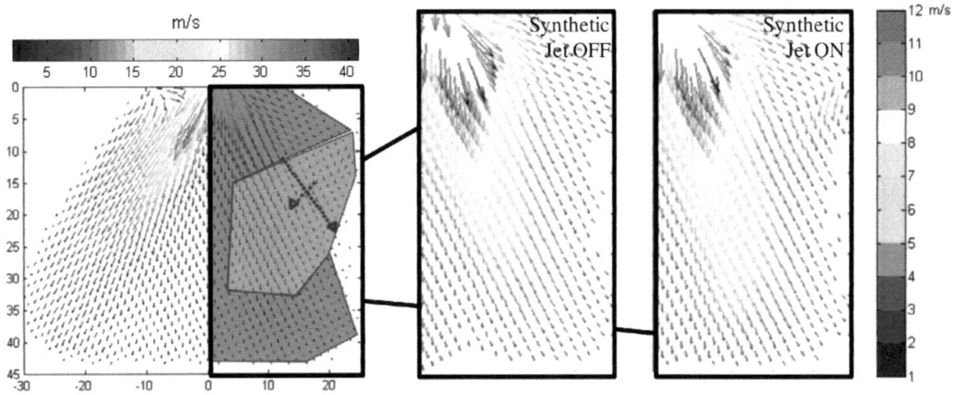

Figure 11. Case 3: Spray velocity field, $(P_{inj} = 5\ MPa, \Delta h = 5\ mm)$.

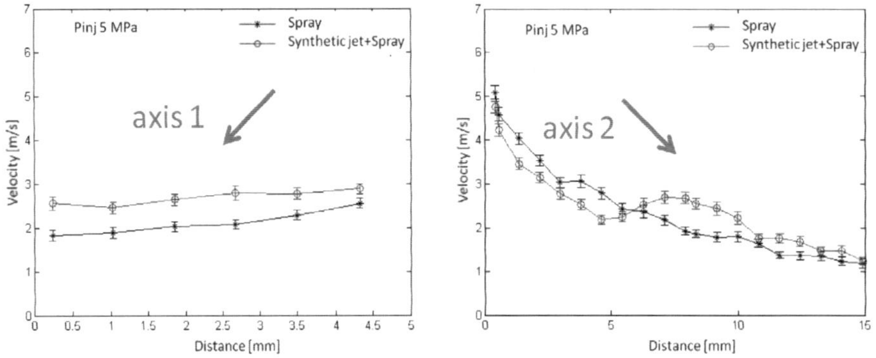

Figure 12. Case 3: Profile plots: longitudinal axis (left) – transversal axis (right), $(P_{inj} = 5\ MPa, \Delta h = 5\ mm)$.

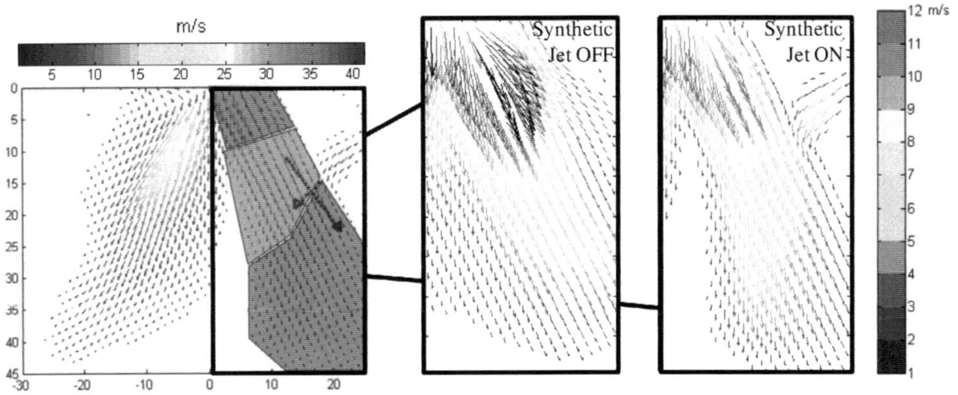

Figure 13. Case 4: Spray velocity field, $(P_{inj} = 10\ MPa, \Delta h = 5\ mm)$.

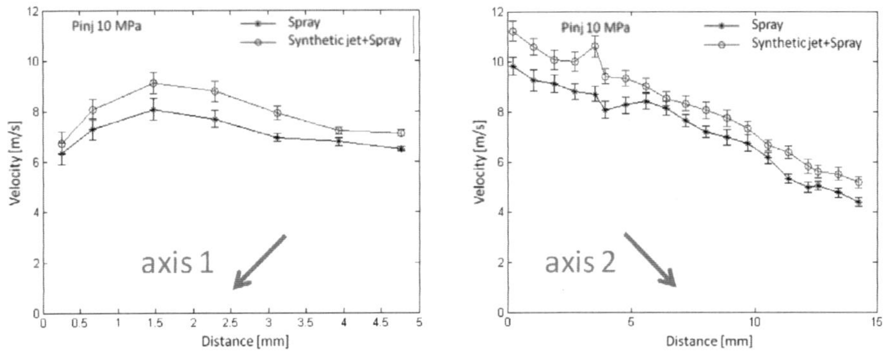

Figure 14. Case 4: Profile plots: longitudinal axis (left) – transversal axis (right), $(P_{inj} = 10\ MPa, \Delta h = 5\ mm)$.

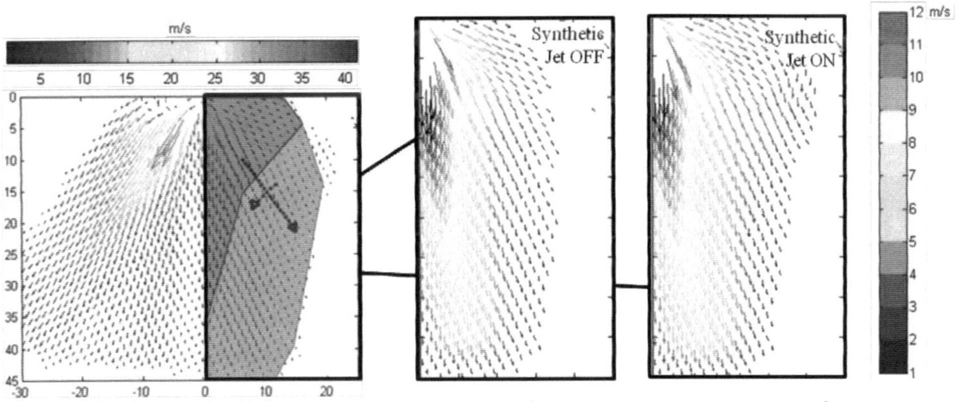

Figure 15. Case 5: Spray velocity field, $(P_{inj} = 5\ MPa, \Delta h = 10\ mm)$.

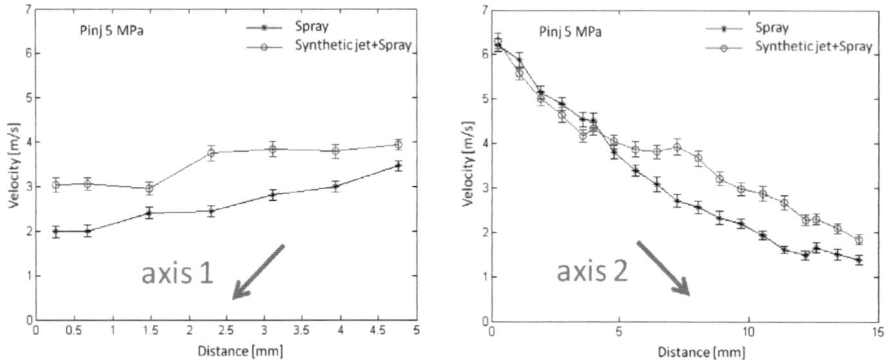

Figure 16. Case 5: Profile plots: longitudinal axis (left) – transversal axis (right), $\left(P_{inj} = 5\ MPa, \Delta h = 10\ mm\right)$.

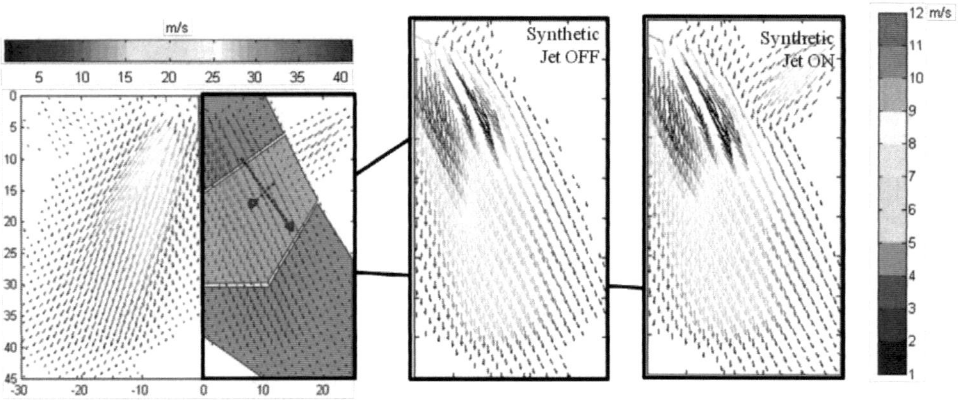

Figure 17. Case 6: Spray velocity field , $\left(P_{inj} = 10\ MPa, \Delta h = 10\ mm\right)$.

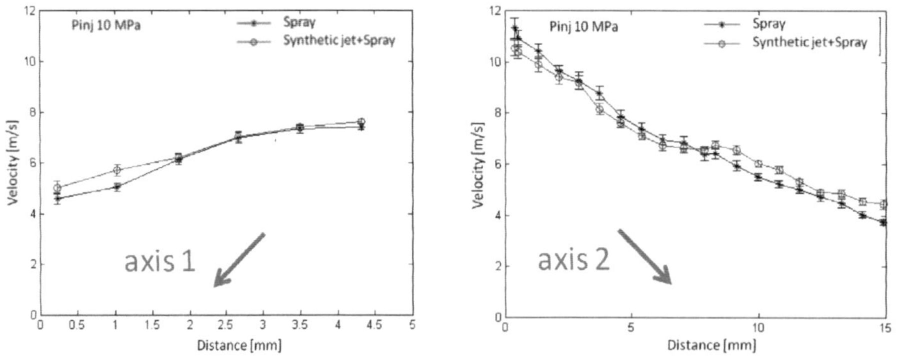

Figure18. Case 6: Profile plots: longitudinal axis (left) – transversal axis (right), $\left(P_{inj} = 10\ MPa, \Delta h = 10\ mm\right)$.

3.2 Spray droplets distribution

PDA measurements have been performed to achieve information about the synthetic jet influence on the atomization process. This technique is able to determine the diameter distribution of the particles which cross the measurement volume, obtaining a local estimation. For this reason it is not possible to characterize the whole spray but the choice of critical and significant points is required. Tests have been carried out for the two injection pressures at Δh=5 mm condition. As for PIV measurements, tests have been repeated with and without the actuator. In particular, 4 points have been set along the longitudinal axis, whereas 7 along the transversal one. The resulted profile plots are reported in Figure 19: the left picture is referred to the longitudinal axis whereas the right one to the transversal one.

As expected the higher injection pressure resulted in an enhanced atomization. Along the longitudinal axis the mean diameter lowers from about 40 µm to less than 35 µm; whereas along the transversal axis values range from about 35 µm to 50 µm undergoing a downward jump of about 10 µm. This is related to the growing of the aerodynamic forces at 10 MPa, which enhances the break-up process.

Looking at the effect of the synthetic jet, an increase in droplets diameter is produced along the longitudinal axis for both pressure conditions. It can be related to a coalescence effect: the particles which move away from the spray periphery are pushed back by the synthetic jet towards to spray axis, colliding with the droplets behind. The synthetic jet is not so efficient to transfer to the particles a high momentum exchange, therefore droplets disintegration is limited and coalescence occurs. This result is in accordance with the work of Pavlova et al. [8].

On the contrary, along the transversal axis the influence of the synthetic jet is almost negligible. Whereas at 10 MPa condition the coalescence effect seems to be still present, at 5 MPa the two curves are practically overlapping.

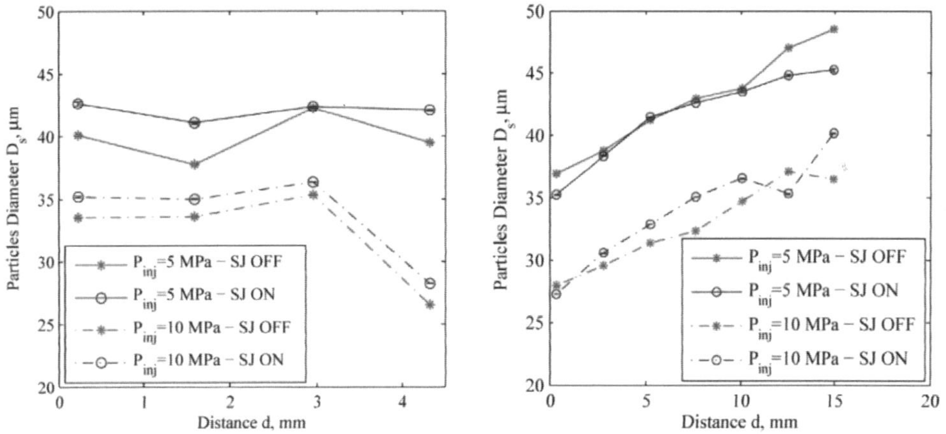

Figure 19. Average droplets diameter (D10) profile plots: longitudinal axis (left) – transversal axis (right).

4. Conclusions

PIV results show that the synthetic jet interacts with the spray locally, energizing the region downstream of the impact and producing higher velocity droplets. This effect tends to reduce increasing the injection pressure, in agreement with the estimated momentum coefficients, due to a high droplets momentum values, which makes their motion alteration more difficult. Moving downward the injector, the air flow impacts against the spray in a region closer to the nozzle. The higher water momentum flux induces a reduction of the air flow-water spray interaction, both in terms of interaction region size and velocity difference.

PDA measurements exhibits a coalescence effect on the droplets distribution along the actuator axis. The influence is higher at the lowest injection pressure. On the contrary, along a transversal axis significant variations are not revealed. This behaviour has likely to be attributed to the relatively long distance between the jet exit orifice and the water spray region where the breakup process is completely accomplished.

Acknowledgements
Prof. Claudio Leone and Dr. Silvio Genna, of the CIRTIBS research centre at the University of Naples "Federico II", must be thanked for their continuous assistance in solving technological issues.
The authors also acknowledge the technical support provided by Mr. Alfredo Mazzei of "Istituto Motori" (CNR) that has been fundamental for the preparation of the equipment, finding smart and rapid solutions.

[1] Cattafesta L N, Sheplak M 2011 Actuators for active flow control *Annual Review of Fluid Mechanics vol.43* pp 247-272.
[2] Pothos S, Longmire E K 2002 Control of a particle-laden jet using a piezo-electric actuator *11th International Symposium on Applications of Laser Techniques to Fluid Mechanics, Lisbon.*
[3] de Luca L, Girfoglio M, Coppola G 2014 Modeling and experimental validation of the frequency response of synthetic jet actuators. *AIAA Journal vol 52* n.8 pp 1733-1748.
[4] Girfoglio M, Greco C S, Chiatto M, de Luca L 2015 Modelling of efficiency of synthetic jet actuators *Sensors and Actuators A: Physical vol 233* pp 512-521.
[5] de Luca L, Girfoglio M, Chiatto M and Coppola G 2016 Scaling properties of resonant cavities driven by piezo-electric actuators *Sensors and Actuators A: Physical vol 247* pp 465–474.
[6] Holman R, Utturkar Y, Mittal R, Smith B L, Cattafesta L 2005 Formation criterion for synthetic jets *AIAA journal vol 43* n 10 pp 2110-2116.
[7] Smith B L, Glezer A 1998 The formation and evolution of synthetic jets *Physics of Fluids (1994-present) vol 10* n.9 pp 2281-2297.
[8] Pavlova A A, Otani K, Amitay M 2008 Active control of sprays using a single synthetic jet actuator *International journal of heat and fluid flow, vol 29* n.1 pp 131-148.
[9] Marchitto L, Valentino G, Merola S S, Tornatore C 2015 Characterization of alcohol sprays from multi-hole injector for DISI engines through PIV technique (No. 2015-01-0927) SAE Technical Paper.
[10] Sepe V 2015 Optical Investigation of a Water Spray Controlled by a Synthetic Jet Device *Master's Degree Dissertation* University of Naples "Federico II".

Development of a background-oriented schlieren technique with telecentric lenses for supersonic flow

F Cozzi[1], E Göttlich[3], L Angelucci[2], V Dossena[1] and A Guardone[2]

[1]Politecnico Milano, Dipartimento di Energia, Via Lambruschini 4, 20156 Milano, Italy
[2]Politecnico di Milano, Dipartimento di Scienze e Tecnologie Aerospaziali, Via Lambruschini 4, 20156 Milano, Italy
[3]Graz University of Technology, Institute for Thermal Turbomachinery and Machine Dynamics, Inffeldgasse 25/A, 8010 Graz, Austria

E-mail: fabio.cozzi@polimi.it

Abstract. Background oriented schlieren (BOS) is a quantitative optical technique which exploits light deflection occurring in non-homogeneous transparent media. It allows to indirectly measure the density gradients by analysing the apparent displacement of features of a background pattern when imaged through the investigated flow. Thanks to its simple set-up and to the consolidated data reduction technique based on cross-correlation algorithms the BOS technique has progressively attracted the interest of the researchers. In this work a BOS system using a telecentric lens system has been set up in order to improve measurement accuracy and to avoid 3D effects arising from using conventional entocentric lenses. The design of the telecentric lens system is reported along with an analysis of its performance in term of spatial resolution. Some preliminary tests on a supersonic flows are also reported.

1. Introduction

The background orientend schlieren (BOS) is a line of sight optical technique that allows quantitative density measurements in inhomogeneous transparent flow. While the technique dates back to the work of Hubert Schardin [1], modern applications of BOS exploiting digital images recording and processing were pioneered in the early 2000's [2]-[3]. Compared to schlieren and shadow techniques BOS is simpler to set-up and allows to get quantitative density data more easily [4]-[5]. The BOS technique exploits light ray deflection occurring in inhomogeneous transparent medium to measure density gradients. Light ray deflection is estimated by comparing the images of a background pattern taken with and without the schlieren object (i.e. an inhomogeneous density field) in between the camera and the pattern. The Gladstone-Dale relationship between refractive index and density allows to retrieve density gradients from the measured displacements.

The popularity of BOS has grown rapidly in the last ten years also thanks to the continuous improving in image acquisition and analysis [6]. The BOS technique has been applied to investigate a wide spectra of phenomena: supersonic flows [4]-[7], rotor blade tip vortex of a full scale helicopter [8], free flight experiment [9], heat transfer in natural convection [10] and free jet [11] to name just but a few. Stereoscopic BOS configurations were used for the localization of a wing-tip vortex in a transonic wind tunnel [12], while tomographic approach have been also applied to single or multicamera BOS set-up to reconstruct the 3D density fields [13]-[14]. The use of natural background for large scale outdoor BOS system have been also analyzed by Hargather and Settles [15]. A more comprehensive overview on applications of the BOS technique can be found in Raffel [6].

Imaging in BOS systems is usually accomplished by using entocentric lenses. The main feature of such lenses is to realize a central projection of the object plane (the subject) into the image plane (the camera sensor); chief rays in entocentric lenses are inclined respect the optical axis and image magnification depends upon object position even within the region of sharp imaging (i.e. the depth of field). Diverging light recording hinder the evaluation of the fluid density mainly for two reasons: firstly, light rays experience a variation in the gradient of the refractive index along their way through the schlieren object; secondly, the location at which ray displacement is measured differs from the location in the schlieren object that generated it. Thus even in plane 2D flows the diverging light recording can add uncertainty to the density gradient measurements.

Improvement in the quantitative application of BOS can be achieved by alleviating the previous issues through the use of telecentric lens. The main feature of such lens systems is that either its entrance pupil or its exit pupil is located at infinity [16],[17]. When both pupils are at infinity the system has an afocal configuration and it is called double side telecentric. The principal rays of a double side telecentric objective are parallel to the optical axis in both the image and the object space. Thus image originates by the parallel projection of the object onto the image plane and magnification is independent from the position of the object and the image planes. Telecentric systems offer also a quite wide depth of field being the aperture size usually quite small as compared to the focal length of the lenses. The most simple double side telecentric configuration is made by two positive lenses, separated by a length equal to the sum of their focal lengths, along with an aperture located at the common focal position of both lenses, see figure 1.

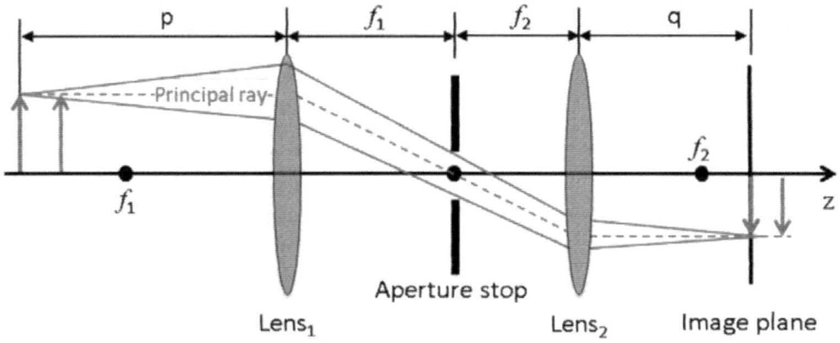

Figure 1. Double side telecentric lenses (afocal system).

When applied to BOS systems, double side telecentric objective is expected to offer some advantage: for plane 2D schlieren object being the imaging rays parallel to the optical axis they cross the density field along lines of constant density gradient, thus quantitative data can be obtained without the need of any correction or tomographic reconstruction. Moreover the location in the image at which the ray deflection is measured corresponds to the physical location at which the deflection is generated. Finally shadows cast by walls or obstacle are sharper if compared to entocentric lenses, thus measurements can be carried out at closer distance. The main disadvantage of telecentric lenses is the need for a frontal lens of at least the same size of the field of view.

Telecentric lenses are commonly employed in metrological application, while just few applications to BOS and PIV systems are reported in the literature [4], [18], [16]. Elsinga et al. [4] used a schlieren system with a diaphragm in place of the knife to perform BOS measurements in a supersonic 2D flow, this optical configuration is akin to a telecentric system. They compared estimated densities and density gradients with those obtained with an entocentric lens system showing better quantitative results for the telecentric configuration.

Recently Ota [18] applied a double telecentric-colored grid background-oriented schlieren (CGBOS) systems to investigate the supersonic flow around both axysimmetric and asymmetric models. The results were compared to those obtained with an entocentric lens system showing some improvement in the measured displacement field.

In this work a telecentric lens system for a BOS system has been set up. The design of the telecentric lens system is reported along with an analysis of its performances. Preliminary tests on a supersonic flows in a plane 2D over-expanded nozzle are also reported and commented.

2. Principles of the background oriented schlieren technique

Background oriented schlieren is an optical technique to measure density gradients within a transparent fluid. In optically transparent fluids the refractive index, n, is related to density, ρ, through the Gladston-Dale relation [19]:

$$n = 1 - G\rho \tag{1}$$

where G is the so-called Gladstone-Dale constant. The Gladstone-Dale constant depends on both the nature of the propagation medium and the radiation wavelength, λ. Furthermore, equation (1) contains implicitly temperature and pressure that can be put in evidence by knowing the equation of state $\rho = \rho(T, p)$. According to equation (1) any density gradient gives rise to refractive index gradient and eventually to light rays deflection [17]. Thus details of a background object show an apparent displacement when seen through an inhomogeneous density field. By comparing the images of the background taken with and without the schlieren object the local image displacement $(\Delta x, \Delta y)$ can be measured. Displacements are converted to refractive index gradients by analyzing the ray path and finally to density gradients through equation (1). Advanced cross-correlation techniques commonly used in PIV measurements [11], or even optical flow algorithms [20], can be used to evaluate the displacement vector $\Delta \vec{r}(x, y) = (\Delta x, \Delta y)$ from BOS images.

A simplified scheme of a BOS setup is reported in figure 2. It is composed by a digital camera, imaging lenses, a background object and a lighting system. The background object usually is a random dot pattern on a surface behind the schlieren object. The pattern preferably has to have the highest spatial frequency that can be imaged with sufficient contrast.

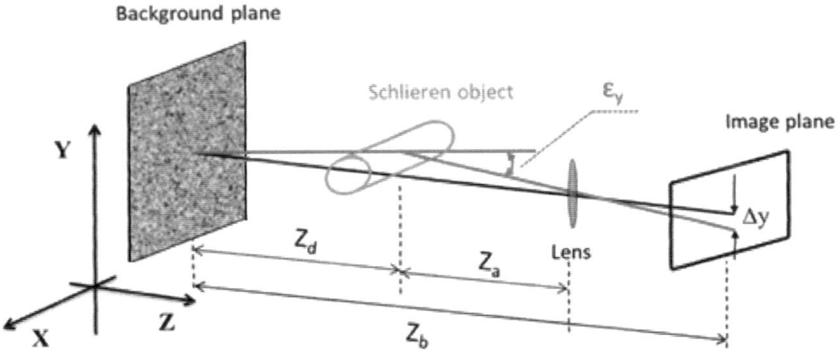

Figure 2. Sketch of a BOS optical setup.

For paraxial recording and small deflection angles ($\varepsilon_y \approx \tan \varepsilon_y$), the apparent image displacement Δy along the y direction is proportional to the deflection angle ε_y [21], figure 2:

$$\Delta y = M Z_d \varepsilon_y \tag{2}$$

where Z_d is distance between the background and the plane at which the light ray deflection can be assumed to have fully occurred and $M = (Z_b - Z_d - Z_a)/(Z_d + Z_a)$ is the magnification of the entocentric lens system. Z_a is the schlieren object to lens distance and Z_b is the background plane to image plane distance. Equation (2) holds also for telecentric lens system provided the magnification M is that given by equation (6) [18]. The sensitivity S of the BOS system is defined as $S = \Delta y/\varepsilon_y = MZ_d$. The deflection angle ε_y and the refractive index gradient along the y direction, $\partial n/\partial y$, are linked through equation (3) [22].

$$\varepsilon_y(X_s, Y_s) = \int_{Z_s}^{Z_s+W} \frac{1}{n}\frac{\partial n}{\partial y} dz \tag{3}$$

where W is the width of the schlieren object and X_s, Y_s and Z_s are the entrance coordinate of the light ray in the schlieren region. According to equation (1) and equation (3) the deflection angles ε_y results to be a linear function of $\partial\rho/\partial y$ in plane 2D density fields, see equation (4),

$$\varepsilon_y(X_s, Y_s) \cong \frac{G}{n_0} W \frac{\partial\rho}{\partial y} \tag{4}$$

where n_0 is the refractive index of the fluid at (X_s, Y_s, Z_s). A similar relationship also holds for displacements and deflection angles along the x direction, thus the density gradient $\vec{\nabla}\rho = (\partial\rho/\partial x, \partial\rho/\partial y)$, can be estimated from the image displacement vector, $\Delta\vec{r}(x_s, y_s) = (\Delta x, \Delta y)$, through equation (5):

$$\vec{\nabla}\rho\left(X_s, Y_s, Z_s + \frac{W}{2}\right) \cong \frac{n_0}{W \cdot G \cdot MZ_d} \Delta\vec{r}(x_s, y_s) \tag{5}$$

where (x_s, y_s) are the coordinates (X_s, Y_s) projected into the image plane.

3. Telecentric lens design and testing

The telecentric systems is made of two positive thin lenses, having focal lenghts f_1 and f_2, separated by a distance equal to the sum of their focal distances. A diaphragm is located in the common focal point, see figure 1. The magnification of the telecentric lenses system, M, and the conjugate image and object distances p and q are given by equation (6) and equation (7), respectively [16].

$$M = \frac{f_2}{f_1} \tag{6}$$

$$q = f_2\left[\frac{f_2}{f_1}\left(1 - \frac{p}{f_1}\right) + 1\right] \tag{7}$$

For a real image to be formed on the camera sensor plane the image to lens distance should be positive, i.e. $q > 0$, thus equations (6) and (7) imply $p \leq f_1(1 + 1/M)$. For magnification not too far from unit a real image is formed at a distance p of the order of f_1. Combining equation (6) and equation (2) the image displacements, Δy, results to be a linear function of ray deflections, ε_y, as per equation (8).

$$\Delta y = \frac{f_2}{f_1} Z_d \varepsilon_y \tag{8}$$

Differently from the entocentric system, Δy is not affected by Z_a. According to Berger [23] when the object to lens distance q is equal to f_1 magnification shows a minimum sensitivity to axial positioning errors of the optical elements. Thus the system was designed for coniugate distances $q = f_1$ and $p = f_2$.

The required lens magnification is set by the camera sensor size, L_c, and the physical field of view, L_{FW} as per equation (9).

$$M = \frac{L_C}{L_{FW}} \tag{9}$$

In our case we use a CMOS camera having a square sensor of size $L_c \cong 16.4$ mm. Setting the field of view equal to the height of the exit section of the nozzle, i.e. L_{FW} =30 mm, the magnification resulted to be about 0.55. The telecentric system was assembled by using two lenses available in the laboratory and suitable for the required magnification. The lenses have focal lengths of f_1=310 mm and f_2=150 mm and diameters of about D_{L1}=80 mm and D_{L2}=51 mm, respectively. From equation (4) the nominal magnification resulted to be M_{nom} = 0.484, this value is about 15% smaller than desired giving a slightly wider field view.

A suitable aperture has to be placed at the common focal point of the two lenses to get the telecentric optical system. The diameter of the aperture, d, besides setting the amount of light reaching the image plane, affects also the depth of field and the resolving power of the optical system. By decreasing d the system became more tolerant to out of focus [23], while depth of field increases and resolving power decreases.

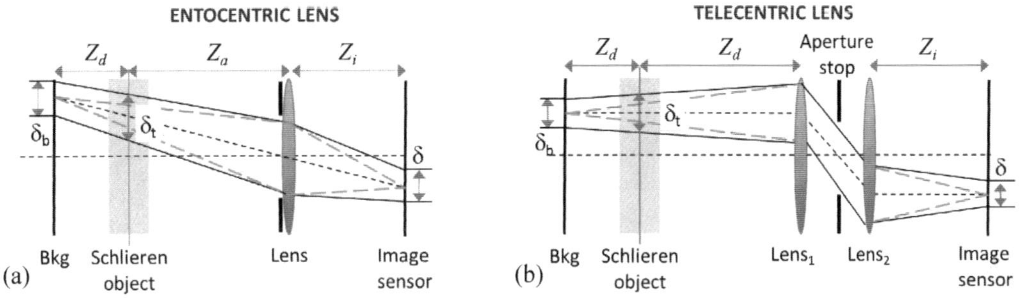

Figure 3. BOS setup for entocentric (a) and telecentric (b) optical systems, $\delta_b = \delta/M$ is the size of the circle of confusion projected back on the object plane.

The optimum size of d is chosen by maximizing the spatial resolution, δ_t, of the BOS measurement. The latter is the size of the schlieren region that contribute to the apparent displacement of the smallest resolvable feature of the background target, thus refractive index variation occurring in region smaller than δ_t cannot be resolved. Figure 3 shows the BOS setups for entocentric and telecentric optical systems along with their spatial resolution δ_t. The analysis of δ_t for entocentric BOS system was examined by Klinge et al. [12] and by Gojani et al. [24], the analysis was later extended to telencentric system by Ota et al. [18]. As done in [24] and in [18] and differently from [12] the spatial resolution δ_t considered here do not take into account the size of the interrogation area used to estimate the apparent image displacement. Expression of δ_t for entocentric and telecentric systems are given in equations (10) and (11), respectively.

$$\delta_t = \frac{M}{M+1} Z_d \frac{d}{f} + \frac{\delta}{M}\left(1 - \frac{Z_d}{Z_d + Z_a}\right) \tag{10}$$

$$\delta_t = Z_d \frac{d}{f_1} + \frac{\delta}{M} \tag{11}$$

Where δ is the diameter of the circle of confusion. The first term on the RHS of both equations (10) and (11) is the diameter at the refractive plane of the light cone emitted from one point on the background, while the second one is the size of the confusion diameter projected onto the refractive plane. It has to be pointed out that lens aberrations are not taken into account, thus equations (10) and (11) are expected to underestimate the value of δ_t.

The size of the circle of confusion depends on the size of the aperture that control the amount of light collected by the optical system [17],[25]. According to the Rayleigh criterion and for monochromatic light of wavelength λ, the diameter of the circle of confusion of a telecentric lens system is $\delta = 1.22\lambda f_2/d$ [25], by substituting δ in equation (11) we get for δ_t the expression in equation (12). The spatial resolution δ_t results to be a function of the aperture size d, while the focal length of the frontal lens, f_1, the distance Z_d and the wavelength λ can be considered as fixed parameters.

$$\delta_t = Z_d \frac{d}{f_1} + \frac{1.22\,\lambda\, f_1}{d} \tag{12}$$

The function $\delta_t(d)$ is plotted in figure 4 for f_1=310 mm, λ=460 nm and Z_d =30 mm, and it shows to have a minimum. The aperture size, d_{opt}, corresponding to that minimum is given by equation (13).

$$d_{opt} = \left(1.22 \frac{\lambda}{Z_d}\right)^{1/2} f_1 \tag{13}$$

For $d = d_{opt}$ the two terms in the LHS of equation (12) result to be equal and equation (14) gives the optimum spatial resolution, δ_{opt}, of the telecentric BOS system.

$$\delta_{opt} = 2 * (1.22\lambda Z_d)^{1/2} \tag{14}$$

It is interesting to note that while d_{opt} is a linearly function of f_1, δ_{opt} does not depend on either f_1 or f_2. This means that for fixed Z_d and λ, f_1 and f_2 can be freely chosen to adjust magnification without affecting δ_{opt}. By using equation (13) Z_d can be eliminated from equation (14) getting $\delta_{opt} = 2.44\,\lambda\, f_1/d_{opt}$, thus optimum resolution and optimum aperture are inversely proportional.

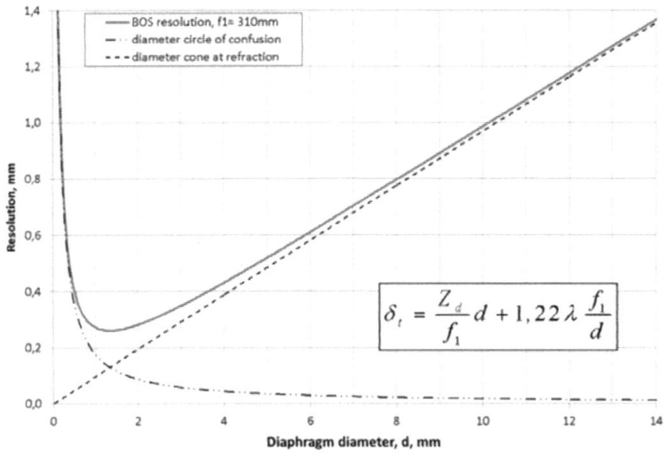

Figure 4. Continuous line: spatial resolution of telecentric BOS systems, δ_t, computed from equation (12), f_1=310 mm, λ=460 nm and Z_d =30 mm. Dashed lines are contribution to δ_t from the two single terms on the RHS of equation (12).

Figure 5 shows d_{opt} and δ_{opt} computed through equation (13) and equation (14) for background to schlieren object distance, Z_d, varying between 0 and 100 mm, while f_1=310 mm, λ=460 nm and $Z_d + Z_a$ =300 mm are kept constant. Figure 5 evidence that d_{opt} decreases and δ_{opt} increases as Z_d increases; thus to improve resolution short background to schlieren object distance, Z_d, and wide aperture, d, are both required. Being the BOS sensitivity, S, decreasing when Z_d decrease a trade-off is required between resolution and sensitivity.

The assembled telecentric objective attached to the CMOS camera is shown in figure 6. A manually variable iris is placed in the common focal point of the two lenses, this allows d to be freely adjustable between 0.8 mm and 25 mm.

IOP Conf. Series: Journal of Physics: Conf. Series **778** (2017) 012006 doi:10.1088/1742-6596/778/1/012006

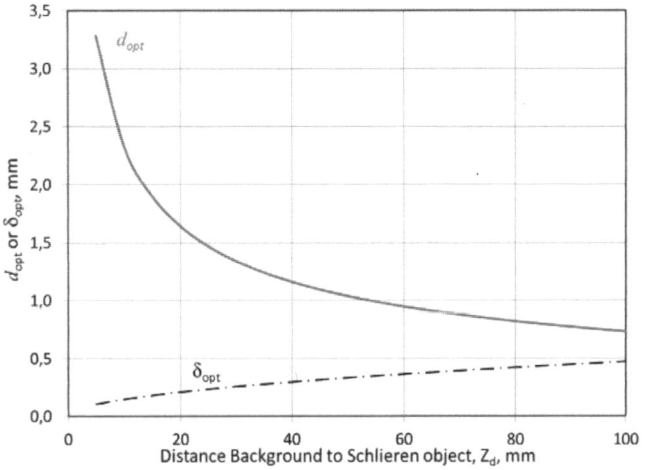

Figure 5. Optimum aperture size, d_{opt}, and optimum resolution δ_{opt} computed through equations (13) and (14) as a function Z_d (f_1=310 mm and λ=460 nm).

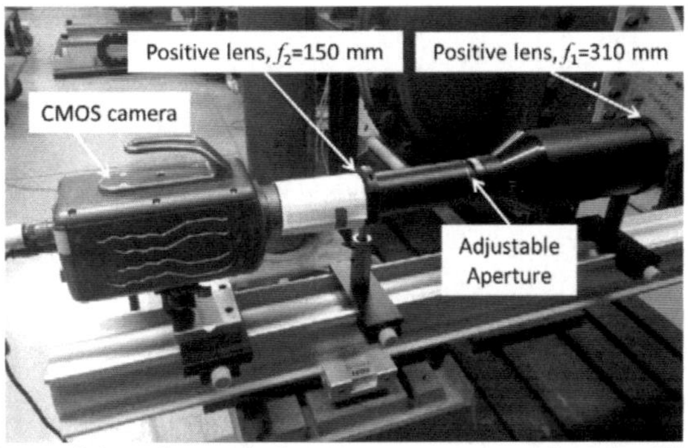

Figure 6. Photo of the telecentric lens systems mounted on the CMOS camera.

The actual magnification of a lens system depends on parameters that cannot be easily measured with good accuracy (i.e. actual position of lenses, their actual focal length etc.) thus calibration is required. The telecentric assembly was tested for magnification and for telecentric degree by imaging a printed calibration pattern (square grid 5 mm by side of circular dots) located at the focusing distance (i.e. 310 mm) and at +/-20 mm away from it (about +/- 6.5 % of f_1). For aperture sizes of 1, 2, and 7 mm, magnification M resulted to be equal to 0.479, i.e. very close to the nominal value of 0.484; magnification changed of about +/- 0.2% over the out of focus rage of 40 mm. Depth of field was also roughly estimated by imaging a ruler 30° inclined respect to the optical axis. The depth of field resulted to be about 65 mm for both $d \cong 1$ mm and $d \cong 2$ mm, and it decreases to about 40 mm for $d \cong 7$ mm.

Based on previous results and on some preliminary tests with $Z_d = 30$ mm the diameter d=2 mm seemed to be the best choice for the diaphragm aperture. This aperture is twice the optimum one, see figure 4 and figure 5, nevertheless it improves light collections and gives sharper images, without significantly affecting the depth of field and telecentric degree.

To check system sensitivity, S, tests were performed with the telecentric BOS system using a positive thin lens having a focal length f_{test}=1000 mm as a schlieren object. The radial image displacement, $(\Delta x^2 + \Delta y^2)^{1/2}$, is a linear function of both radial distance, r, from the optical axis and sensitivity, as given by equation (15).

$$(\Delta x^2 + \Delta y^2)^{1/2} = S\, r / f_{test} \tag{15}$$

Two values of Z_d were tested: 31.5 mm and 61 mm, for M=0.479 the nominal sensitivity $S = MZ_d$ resulted to be 15.1 and 29.5 mm/rad, respectively. The horizontal image displacement, Δx, along the horizontal axis is shown in figure 7(a), as expected it shows an almost perfect linear variation with r. A linear fit of the measured displacements allows to estimate a sensitivity of 15.2 mm/rad and of 29.5 mm/rad for Z_d=31.5 mm and Z_d= 61 mm, respectively, almost in perfect agreement with the nominal ones. The radial displacement over the whole test lens is shown in figure 7(b) and it evidences a very good axisymmetric distribution. The almost uniform zero-displacement region in the bottom of figure 7(b) is originated by the image of the lens support.

The measured performances evidence the proper design and assembly of the optical system.

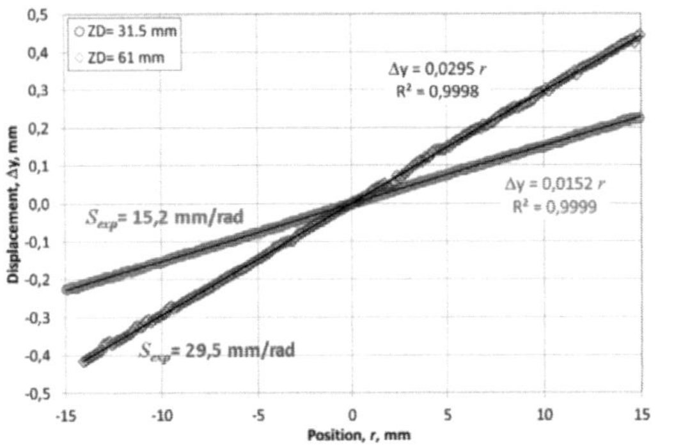

(a)

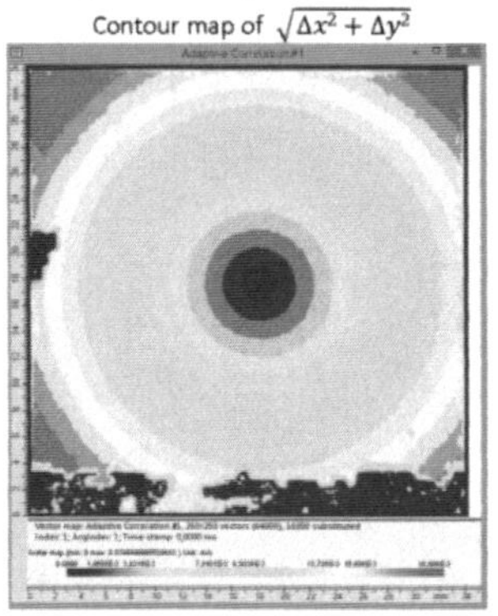

(b)

Figure 7. (a) Horizontal image displacement along the horizontal axis for Z_d= 31.5 mm and 61 mm. (b) Contour of radial displacement $(\Delta x^2 + \Delta y^2)^{1/2}$ for Z_d = 61 mm.

4. Experimental set-up

To test the telecentric BOS system, the flow at the exit of an over-expanded convergent-divergent (CD) nozzle is investigated. Experiments are carried out in a supersonic 2D blow-down wind tunnel. A plane parallel-walled CD nozzle with a nominal area ratio A_e/A_t =1.5 and 12.3 mm in width is attached to the plenum of the wind tunnel through a trumpet-shaped convergent section. The nozzle exit section has a height of about 30 mm. The plenum is a pressurized cylindrical vessel providing the stagnation condition for the flow, a piezoresistive pressure transducer with a full scale of 800 kPa (extended accuracy = 0.08% of the full scale) and a thermocouple probe (<1% accuracy) measure the stagnation pressure and temperature, respectively. Tests are carried out at total pressures of about 350 kPa and 500 kPa and at a constant total temperature of about 291 K (+/0.5 K). The nozzle discharges into ambient atmosphere where pressure is about 1 bar. Two plane BK7 8 mm thick glass windows (refractive index $n_{BK7} \cong 1.52$) provide the transparent side walls for the 2D CD nozzle. A photo of the experimental set-up is shown in figure 8.

Pressure measurements are taken at discrete position along the nozzle axis, pressure taps are located on a brass plate which is put in place of the rear glass window. Due to the brass plate, BOS and pressure measurements cannot not be carried out simultaneously. The plate has 17 pressure taps drilled along its axis of symmetry, taps are 0.5 mm in diameter and equally spaced under 5 mm. Each pressure tap is connected to piezoresistive pressure transducers through silicon pipes, the transducers have a full scale of 350 kPa their extend accuracy was 0.05 % of the full scale. A complete scan requires a few tens of seconds, for each test condition the scan is repeated 3 times. The three measurements series are averaged together and pressures vary of about 0.2 % from scan to scan.

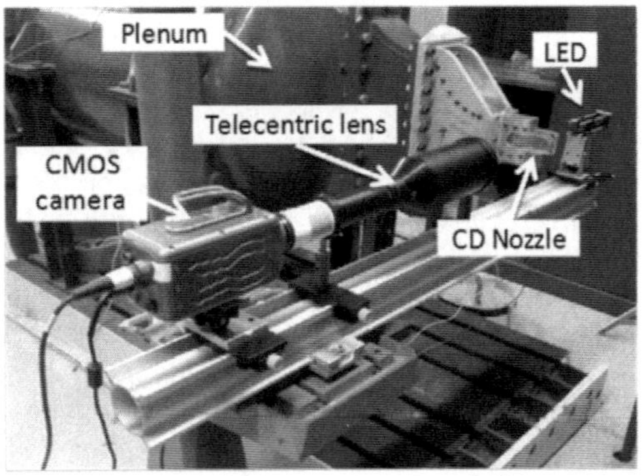

Figure 8. Experimental set-up.

The BOS system is composed by a CMOS camera, the telecentric lens system described above and a blue LED for lighting, see figure 8. The camera has a resolution of 1024*1024 pixels and a pixel pitch of 16 μm. The telecentric lens system uses an aperture size set to 2 mm for all the tests reported here. The lighting system is composed by a high power LED (Osram Ostar LE B Q9WN), emitting at a wavelength of 460 nm. The emitting surface has an area of 0.9*0.9 mm^2 and the LED operates at a constant current of about 700 mA corresponding to a nominal light output power of about 1 W. To improve light uniformity over the translucent target, the central portion of the light beam is expanded by means of a lens system. The recorded images show a spatially uniform and constant in time light intensity.

(a) 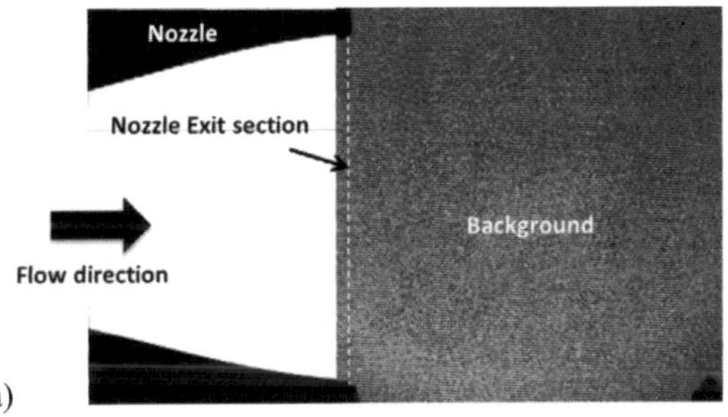

(b) 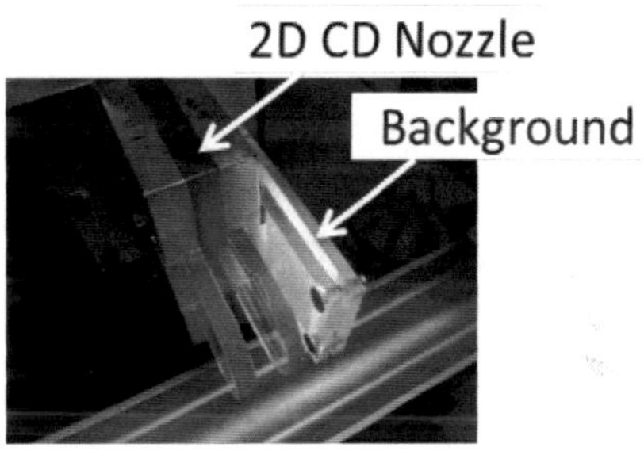

Figure 9. (a) Sample background image recorded by the BOS system, the sketch added to the left of the image shows flow direction and the shape of the diverging section of the nozzle. (b) Background pattern lightened by the LED.

The background is an artificially generated speckle pattern. The spatial distribution of the grey and white structures of the picture used as background is casual and generated using a Gabor noise. This particular type of noise, based on the Gabor's convolution, is particularly indicated for this kind of application because it permits a strict control of the spatial frequency of the generated pattern [26]. The image was printed on a transparent at resolution of 600 DPI (about 42 μm per printer dot) by using a laser printer and a sample background image is shown in figure 9(a). The speckle dots size, estimated from a visual inspection, resulted to be about 4 pixels, this value corresponds also to the width of the cross-correlation peak given by the PIV software. Such dot size is slightly big but still in the optimal range of 2-4 pixels suggested by Adrian [27]. Due to limitation in printer resolution, the speckle dots size couldn't be reduced and the printed image has a more or less regular pattern of small black and white dots with a spatial frequency of about 1/8 pixels^{-1}. The background target is placed between two PMMA plates, each 6 mm thick (refractive index n_{PMMA} ≅1.49), and is firmly stuck to the nozzle structure at 30 mm distance from the rear window, see figure 9(b).

IOP Conf. Series: Journal of Physics: Conf Series **778** (2017) 012006 doi:10.1088/1742-6596/778/1/012006

5. Analysis procedure of BOS images

This section provides a brief description on the technique used for analyzing BOS images. To evaluate the actual spatial position at which the ray displacement originates, the flow images are used as the references ones against which the undistorted background image is compared [4]. The virtual image displacement is estimated by applying an adaptive cross-correlation algorithm (3 steps) with a final interrogation area (IA) of 16 by 16 pixels and 75% overlap, the final mesh size in the object space is about 0.28 x 0.28 mm^2. The analysis is implemented using a commercial PIV software (DynamicStudio 4.15a by Dantec). No image pre-processing is implemented and a single couple of image is analyzed for each run.

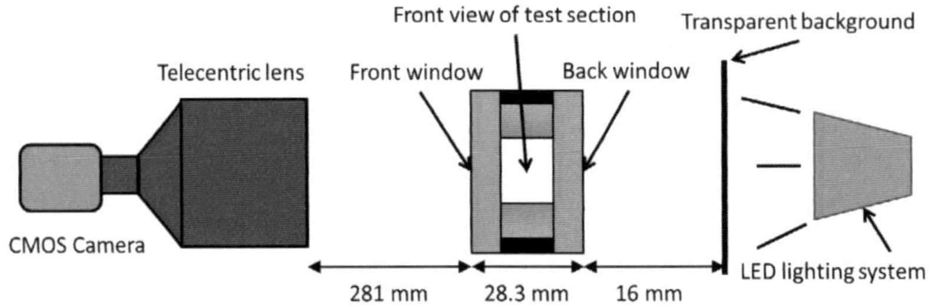

Figure 10. Side view of the telecentric BOS system, including test section (front view) and lighting system.

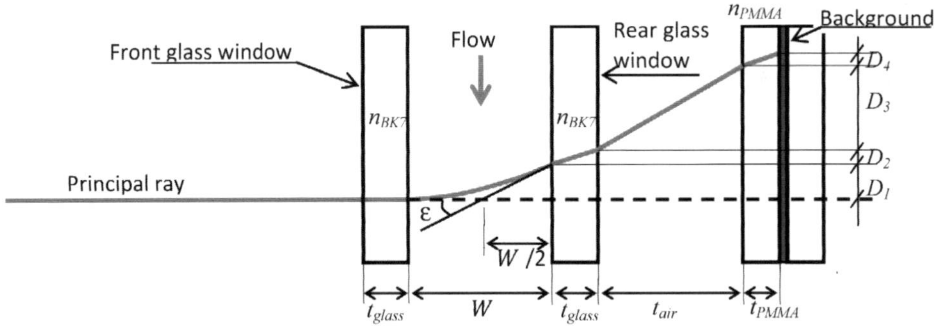

Figure 11. Top view of the test section (the nozzle is not shown) and of the principal ray (continuos thick line). The path of the principal ray in case of no refraction is shown as a dashed line.

In order to recover the angular deflection ε_y from the measured displacement Δy, the schlieren object to background distance Z_d has to be evaluated. Optical components distances are shown in figure 10, while a detailed view of the principal light ray path is shown in figure 11. The back glass window and the PMMA plate contribute to ray refraction, so they have to be taken into account to correctly estimate the angular deflection originated by the schlieren object. Moreover since the nozzle width is comparable to the distance between the background and the back window, it cannot be neglected. For a supersonic wind tunnel with thick windows Elsinga et al. estimated Z_d by using a ray tracing procedure [4], anyhow due to the different optical set-up their results cannot be directly used here. The telecentric lens systems and its wide depth of field allows us to limit our analysis to the principal ray and on how it is refracted by the flow and by the interfaces it encounters along its way, figure 11. The total apparent displacement, $\Delta Y = \sum_i D_i = \Delta y/M$, in the background plane is given by the sum of the displacements occurring in each transparent medium, D_i, see figure 11. Being perpendicular to the principal ray the front window doesn't give any contribution to the total displacement and it is not taken into account. By summing all displacements and assuming small ray deflection angles, i.e. $\varepsilon_y \cong \tan(\varepsilon_y)$, ε_y is given by equation (16),

$$\varepsilon_y \cong \tan(\varepsilon_y) = \frac{\Delta y}{M}\left[\left(\frac{W}{2} + t_{air} + \frac{t_{glass}}{n_{BK7}} + \frac{t_{PMMA}}{n_{PMMA}}\right)^{-1}\right] \quad (16)$$

thus Z_d is given by equation (17):

$$Z_d = \left(\frac{W}{2} + t_{air} + \frac{t_{glass}}{n_{BK7}} + \frac{t_{PMMA}}{n_{PMMA}}\right) \quad (17)$$

where t_{air}, t_{glass} and t_{PMMA} are the thicknesses of the different media shown in figure 11. If we assume that the background is stuck to the outside of the rear window (i.e. t_{PMMA}=0 and t_{air} = 0) equation (16) and equation (17) exactly recovers the ones obtained by Elsinga et al. [4].

Using W =12.3 mm, t_{air}=16 mm, t_{glass}=8 mm and t_{PMMA}=6 mm, we get Z_d = 31.4 mm, Without properly accounting for refractions, we would get $Z_d = (W/2 + t_{air} + t_{glass} + t_{PMMA})$=36.2 mm, i.e. a value almost 15% higher than the correct one, thus angular deflections and density gradients would have been underestimated of almost the same percentage.

A source of error often neglected in BOS measurement is due to vibrations or channel movement occurring during the tests. The unwanted camera or wind tunnel movements add to the ones originated by the phase object and they act as a spurious displacement affecting the estimation of the density field [4]. A proper way to remove the bias is required if accurate density measurement are sought.

Neither density gradient nor density are evaluated in this work, thus bias from channel movement is not of concern, while equation (8) and equation (17) have not to be used. Anyhow due to their relevance for density estimations they have been reported and discussed.

6. Experimental Results

For a fixed nozzle geometry and working fluid, the flow field depends on the nozzle pressure ratio NPR (i.e. the ratio between reservoir total pressure P_0 and external pressure p_e) [28]. If NPR is equal to the design value (NPR_d), the static pressure at the nozzle exit is equal to ambient pressure and the flow inside the nozzle can be assumed iso-entropic. When NPR is greater than NPR_d the nozzle is under-expanded, the exhaust pressure is greater than the outside ones and Prandtl-Mayer fans appear outside the nozzle. On the other hand, if NPR is smaller than NPR_d, the nozzle is over-expanded and some shock waves appear inside or outside the nozzle depending of the over-expansion level. The actual area ratio is difficult to be estimated accurately, thus actual NPR_d is estimated as P_0/p_s where p_s is the static pressure measured at the nozzle exit section when neither shock waves nor flow separation occur in the nozzle. Assuming the exit section of the nozzle located approximately halfway between two pressure taps, p_s is computed as the average of the values measured at those two taps. NPR_d results to be about 5.36, while tests are conducted with NPR=5 and 3.5, thus the nozzle always operates in over-expanded conditions. Boundary layer thickness and uncertainty in the measurement of the area ratio A_e/A_t prevent the accurate estimate of the exit Mach number, $M_{e,iso}$, directly from the latter. For iso-entropic nozzle flow, a more reliable estimate of $M_{e,iso}$ is obtained from pressure measurements by using equation (18),

$$M_{e,iso} = \left\{\frac{2}{\gamma-1}\left[\left(\frac{P_0}{p_s}\right)^{\frac{\gamma-1}{\gamma}} - 1\right]\right\}^{1/2} \quad (18)$$

where the ratio between the specific heats is γ=1.4. For all the tested conditions neither shock waves nor flow separation occurred inside the nozzle, thus iso-entropic flow condition can be assumed. By using equation (18) and the measured pressure ratios, $P_0/p_s \cong 0.187$, we get $M_{e,iso} \cong 1.75$. The exit Mach number estimated from the nominal area ratio, A_e/A_t =1.5, is 1.85, i.e. about 6% higher than the actual one.

When NPR=5, the nozzle is over-expanded and two oblique shock waves appear outside the CD nozzle. The contour maps of horizontal, Δx, and vertical, Δy, displacements are shown in figure 12(a)

IOP Conf. Series: Journal of Physics: Conf. Series **778** (2017) 012006 doi:10.1088/1742-6596/778/1/012006

and 12(b), respectively. A bias of about 7-8 pixels due to wind tunnel/camera movement can be seen in the horizontal displacements, while a much smaller one, about 1 pixels, affects the vertical ones. The oblique lines seen in figures 12(a)-(b) are the traces of the shock waves originated in the upper and lower side of the nozzle by the over-expanded condition. It can be observed that a shock wave is composed by two closely spaced lines, the latter corresponds to regions where the measured displacement is either smaller or bigger than the ones due to the bias. Such behavior is likely due to limitation of the BOS technique in reproducing very sharp density variation and not to the flow, anyhow such issue is behind the scope of the present work and was not further investigated. The over-expansion level is weak enough for a simple reflection to occur where the two shocks waves intersect. The incidence angles of the original as well as of the reflected shocks and the inclination of the shear layer measured to the horizontal direction are marked as θ_1, θ_2 and δ, respectively. They can be computed by means of oblique shock tables using the estimated $M_{e,iso}$, and experimentally measured by a visual analysis of the displacements maps. The computed angles are $\theta_1 \cong 36.7°$,$\theta_2 \cong 36.4°$ and $\delta \cong 2°$, resulting in very good agreement with the measured ones: $\theta_1^{exp} \cong 36°$,$\theta_2^{exp} \cong 36°$ and $\delta^{exp} \cong 2°$.

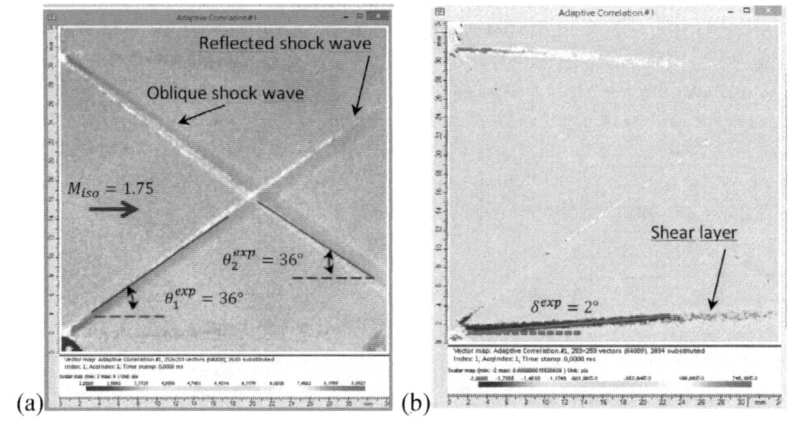

Figure 12. Displacements Δx (a) and Δy (b) in pixel for NPR=5 with P_0=350 kPa, p_e=100 kPa, T_0=294 K and $M_{e,iso}$=1.75.

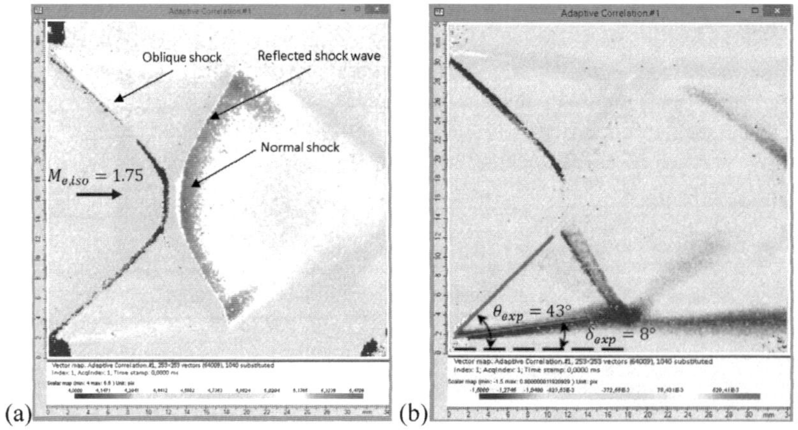

Figure 13. Displacements Δx (a) and Δy (b) in pixel for NPR=3.5 with P_0=350 kPa, p_e=100 kPa, T_0=294 K and $M_{e,iso}$=1.75.

As the NPR decreases the turning imposed to the flow by the oblique shock system increases and a simple shock reflection is no longer possible. As shown in figure 13 the BOS technique is able to catch the complex shock structure named Mach reflection appearing into the flow at NPR=3.5. Three main elements can be seen: an incident oblique shock, a reflected oblique shock and an almost normal shock generally known as Mach steam. The same procedures described above were used to evaluate the slopes of shocks and of the shear layer. The theoretical results are: $\theta_1 \cong 44.2°$, $\delta \cong 8.7°$, they are in good agreement with the measured angles, $\theta_1^{exp} \cong 43°$, $\delta^{exp} \cong 8°$. The above results evidence the capability of the telecentric BOS system to correctly identify shock structures in over-expanded nozzle flows.

7. Conclusions

In this work a BOS technique using a telecentric lens system has been set up in order to improve accuracy and to avoid the 3D effects arising from using conventional entocentric lenses. The design of the telecentric lens system is discussed and reported along with an analysis of the spatial resolution of a BOS system using such kind of lenses. By neglecting lens aberration, the optimized telecentric BOS system has a spatial resolution which depends only on the schlieren object to background distance and on light wavelength, while optimum aperture results to linearly increase with the focal length of the frontal lens.

Based on the described optical design a double telecentric optical system was assembled and tested. The lens system showed good telecentricity, with a magnification changing of less than 0.2 % when the object is displaced of +/-20 mm respect to the sharp focus position.

A formula to relate the measured displacements and the integrated density gradient is inferred for the telecentric optical set-up. To correctly evaluate both the sensitivity of the BOS technique and the angular deflection of the light rays, the schlieren object to background distance has to be properly estimated. To this aim a method to estimate Z_d based on ray-tracing that takes into account the presence of different refractive media is also described.

Finally the telecentric BOS system is used to analyze the supersonic flow outside a convergent-divergent nozzle operating under over-expanded conditions. The analysis of BOS images based on a cross-correlation algorithm allowed to visualize the shock structures and shear layers occurring in the transparent flow fields. For all the investigated cases information about the slopes of the shock waves and shear layers inferred from the measured virtual displacements showed a very good agreement with those evaluated from the tables of oblique shock.

The results confirmed that the designed telecentric lens is suitable for use in a BOS system, while further tests are required to quantitatively assess the improvement achievable in measuring plane 2D density field with telecentric imaging as compared to entocentric ones.

Acknowledgments

This research is supported by ERC Consolidator Grant N. 617603, Project NSHOCK, funded under the FP7-IDEAS-ERC scheme. The authors would also like to thank OSRAM SpA Opto Semiconductors to have provided the high power LEDs and particularly Mr. Gabriele Giaffreda for his kindly assistance.

References

[1] Schardin H 1942 Die Schlierenverfahren und ihre anwendungen *Ergebnisse der exakten Naturwissenschaften* 303
[2] Meier G E A 2000 Hintergrund Schlierenmessverfahren *German Patent, DE 199 42 856 A1*
[3] Raffel M, Richard H and Meier G E A 2000 On the applicability of background oriented optical tomography for large scale aerodynamic investigations *Exp. Fluids* **28** 477–81
[4] Elsinga G E, Van Oudheusden B W, Scarano F and Watt D W 2004 Assessment and application of quantitative schlieren methods: Calibrated color schlieren and background oriented schlieren *Exp. Fluids* **36** 309–25

[5] Venkatakrishnan L and Meier G E a. 2004 Density measurements using the Background Oriented Schlieren technique *Exp. Fluids* **37** 237–47
[6] Raffel M 2015 Background-oriented schlieren (BOS) techniques *Exp. Fluids* **56** 1–17
[7] Clem M M, Zaman K B M Q, Fagan A F and Glenn N 2012 Background Oriented Schlieren Applied to Study Shock Spacing in a Screeching Circular Jet *50th AIAA Aerosp. Sci. Meet.* 1–12
[8] Kindler K, Goldhahn E, Leopold F and Raffel M 2007 Recent developments in background oriented Schlieren methods for rotor blade tip vortex measurements *Exp. Fluids* **43** 233–40
[9] Bauknecht A, Merz C B, Landolt A, Meier A H and Raffel M 2013 Blade tip vortex detection in maneuvering flight using the Background Oriented Schlieren (BOS) technique *AHS 69th Annu. Forum Technol. Disp. "Advancing Vert. Flight Technol. Demanding Environ.* 1–13
[10] Cozzi F, Colombo L P M, Lucchini A, Coghe A, Muzzio A, Pacini F 2010 Background oriented schlieren characterization of the thermal boundary layer over a vertical heated plate in free convection *XVIII AIVELA National Meeting* (Rome) p 6
[11] Richard H and Raffel M 2001 Principle and applications of the background oriented schlieren (BOS) method *Meas. Sci. Technol.* **12** 1576–85
[12] Klinge F, Kirmse T and Kompenhans J 2003 Application of Quantitative Background Oriented Schlieren (BOS): Investigation of a Wing Tip Vortex in a Transonic Wind Tunnel *Proc. PSFVIP4* 1–11
[13] Goldhahn E and Seume J 2007 The background oriented schlieren technique: Sensitivity, accuracy, resolution and application to a three-dimensional density field *Exp. Fluids* **43** 241–9
[14] Venkatakrishnan L and Suriyanarayanan P 2009 Density field of supersonic separated flow past an afterbody nozzle using tomographic reconstruction of BOS data *Exp. Fluids* **47** 463–73
[15] Hargather M J and Settles G S 2010 Natural-background-oriented schlieren imaging *Exp. Fluids* **48** 59–68
[16] Konrath R and Schröder W 2002 Telecentric lenses for imaging in particle image velocimetry: a new stereoscopic approach *Exp. Fluids* **33** 703–8
[17] Born M and Wolf E 1999 *Principles of optics: Electromagnetic Theory of Propagation, Interference and Diffraction of Light* (Cambridge, New York: Cambridge University Press)
[18] Ota M, Leopold F, Noda R M K 2015 Improvement in spatial resolution of background-oriented schlieren technique by introducing a telecentric optical system and its application to supersonic flow *Exp. Fluids* **56** 48
[19] Vasil′ ev L A 1971 *Schlieren methods* (New York: Israel Program for Scientific Translations)
[20] Atcheson B, Heidrich W and Ihrke I 2009 An evaluation of optical flow algorithms for background oriented schlieren imaging *Exp. Fluids* **46** 467–76
[21] Raffel M, Willert C and Kompenhans J 2007 *Particle Image Velocimetry* (Berlin, Heidelberg: Springer Berlin Heidelberg)
[22] Merzkirch W 1987 *Flow visualization* (Orlando, Florida: Academic Press, Inc.)
[23] Berger C 2002 Design of telecentric imaging systems for noncontact velocity sensors *Opt. Eng.* **41** 2599
[24] Gojani A B, Kamishi B and Obayashi S 2013 Measurement sensitivity and resolution for background oriented schlieren during image recording *J. Vis.* **16** 201–7
[25] Gaskill J D 1978 *Linear systems, Fourier transforms, and optics." Linear Systems, Fourier Transforms, and Optics* (New York: John Wiley and Sons)
[26] Lagae A, Lefebvre S, Drettakis G and Dutré P 2009 Procedural noise using sparse Gabor convolution *ACM Trans. Graph.* **28** 1
[27] Adrian R J and Westerweel J 2011 *Particle Image Velocimetry*
[28] Thompson P A 1972 *Compressible-fluid dynamics* (New York: McGraw-Hill)

IOP Conf. Series: Journal of Physics: Conf. Series **778** (2017) 012007 doi:10.1088/1742-6596/778/1/012007

Feasibility study of a vision system for on-line monitoring of rolling contact fatigue tests

I Bodini[1], G Sansoni[2], M Lancini[1], S Pasinetti[2], and F Docchio[1]

[1]Dept. of Mechanical and Industrial Engineering, University of Brescia, via Branze, 38, 25123, Brescia, Italy
[2]Dept. of Information Engineering, University of Brescia, via Branze, 38, 25123, Brescia, Italy

ileana.bodini@unibs.it

Abstract. Wear and Rolling Contact Fatigue (RCF) tests on wheel/rail specimens are important to develop wheels of new materials for improved lifetime and performance, able to operate in harsh environments and at high rolling speeds. We have studied the feasibility of a novel non-invasive all-optical system, based on a high-speed video camera and two laser illumination sources, which is able to continuously monitor the dynamics of the specimens used to test wheel and rail materials, in a laboratory test bench. Surface micro- and macro-topography are monitored using blob analysis and 3D laser triangulation respectively. Blob analysis yields to good discrimination among the specimens, in terms of wear induced surface damage; the 3D measurement, which is characterized by a resolution of 0.033 mm, is able to monitor RCF effects.
The system is described with the aid of end-cycle specimens, as well as of intermediate specimens, prior to its installation in the test bench for rolling contact tests.

1. Introduction

Monitoring wheels and rails is important to increase safety in transportation [1, 2]. This can be done on-site, using different non-destructive inspection techniques such as electromagnetic tomography [3], axle box acceleration measurements [4], vision systems mounted on trains to detect rail conditions [5-7], and in laboratory, on test benches performing rolling contact fatigue tests [8-11].
Laboratory monitoring of damage in rolling contact can be effectively performed using optical techniques [12-16]. However, the rotation speeds of the specimens in current test benches, which is in the order of 500 rounds per minute (rpm), prevent the use of conventional photo cameras or video cameras without stopping the specimens, or at least without slowing them down, with the risk of altering the wear/fatigue process.
Recently, the use of high-speed video cameras, able to acquire images at very high frame-rate is becoming customary in a number of applications (from automotive, to industry, inspection, etc.). A recent paper [17] documents the use of these cameras to monitor a rolling-contact fatigue test for polymeric materials, illuminated by incoherent light. This approach is promising, since the high frame rate allows the monitoring of the specimens without reducing their speed, and without appreciable blurring, provided that the frame exposure is short enough, and, as a consequence, that the illumination is appropriate.
In addition to simple imaging with incoherent light, the use of lasers to monitor roughness as well as three-dimensional characteristics of surfaces is well known. In fact, the measurement of roughness (in

this case, generally related to wear effects) through diffuse scattering from the surface can be combined to the measurement of possible alterations of the 3D profiles of the specimen surface by means of suitable 3D techniques based on laser stripes [18]. Micro- and macro topological measurements can thus be effectively combined to the high-speed frame acquisition when a diffuse laser beam and a laser stripe are used as the illumination source for the specimen during the test, replacing incoherent light sources. The high laser brilliance makes these sources ideal also in the case of very short (tens of microseconds) exposure times.
In a companion Laboratory of our Department a detailed analysis of rolling contact fatigue on wheel-rail pairs is underway, in the framework of an industrial research aimed at improving the wheel performance and lifetime [19-22]. Dry and wet tests are conducted to simulate the different environmental conditions for train transportation. To further improve the test bench, the development of an on-line vision-based apparatus, able to work at a convenient distance from the test bench, and at variable specimen speeds, was requested.
This paper reports the results of a feasibility study carried out on a laboratory prototype based on a high-speed video camera for the wear monitoring of disk shaped specimens. In this activity, special care has been devoted to the design of efficient illumination and of dedicated processing procedures for the monitoring of the surface variations due to both wear and contact rolling fatigue (RCF) damages.

2. Experimental set-up

2.1. Optical layout of the system

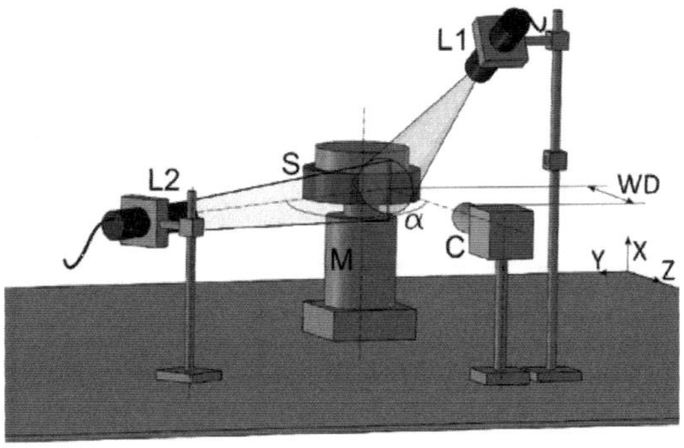

Figure 1. Layout of the laboratory experimental set-up. S: specimen, M: motor, C: high-speed camera, L1: defocused laser, L2: 3D laser stripe.

The schematic layout of the vision system for the cyclic contact experiments is depicted in Figure 1. The measurement target is a disk-shaped specimen S, fixed to the axle of a stepper motor M (HY200 2220 0100 AX08 – SCT Servo Control Technology), with a step angle of 1.80° ± 0.09°. The specimen can rotate at variable speeds up to f_s= 500 rpm, corresponding to 8.3 Hz. The heart of the observation system is the high-speed camera C (PROMON 501 – AOS Technologies AG), which can operate at increasing frame rates (f_c), by correspondingly decreasing the frame area. As an example, at 85 fps maximum frame area is (2048 × 1088) px, whereas at 3391 fps, it is reduced to (320 × 96) px. The exposure time is software-adjustable from 13 μs to the reciprocal of the camera acquisition frequency. To ensure optimal acquisition of the specimens under the operating rotational speed of 500 rpm, the frame rate is set at f_c = 377 fps, with an exposure time equal to 40 μs. The corresponding image resolution is (1280 × 240) px. The camera is equipped with a f/50 mm lens, and placed at a working distance WD of 300 mm from the specimen. The field of view is 26 mm along the X coordinate and 5 mm along the Y coordinate and the resulting spatial resolutions are Rs_X = 0.020 mm/px and

R_{SY} = 0.021 mm/px. At the image plane, X and Y are parallel to the image rows and columns respectively.

Figure 2. System illumination: (a) pattern projected by source L1, (b) pattern projected by source L2. The space coordinates X and Y are shown.

The camera can operate in free-running mode: the frame acquisition runs freely, and all the acquired frames are stored in a cyclic buffer within the Hard Disk of the PC. When the buffer is full, any new frame is inserted in, at the expense of the oldest acquired frame. The camera can also work in the trigger mode: in the event of a trigger signal, the camera stores the event together with the recorded frames.

2.2. The illumination sources

Monitoring the surface of rotating specimens at high speeds requires very low exposure times, to avoid image blurring. At speeds of 500 rpm, the exposure time is set equal to 40 μs. To enhance the visibility of the surface features, coherent light source is used: the high laser brilliance, combined with the possibility of shaping the irradiated areas, allows us to obtain suitable illumination of the specimens for all measurement purposes.

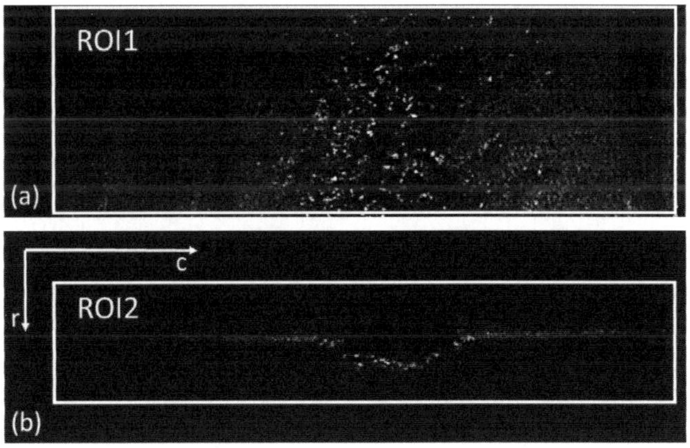

Figure 3. Acquired images from the laser sources. (a) Diffused light from L1 (frame ROI1); (b) Deformed light pattern from L2 (frame ROI2). Parameters r and c correspond to image rows and columns respectively.

In the presented system, two semiconductor laser modules (Lasiris^{TM} SNF Laser), emitting 10 mW at 670 nm, have been assembled: they are denoted as L1 and L2 in Figure 1. The illumination pattern of

each source is shown in Figure 2; the corresponding acquired images are visible in Figure 3. Laser L1 projects a suitably defocused beam onto the specimen surface (Figure 2.a), and the scattered light carries information about the surface dips and crests (Figure 3.a) related to the spalling of the material, due to wear and fatigue effects. Laser L2 is equipped with a cylindrical lens that projects a line pattern (Figure 2.b), and that is mounted at an angle α of about 60° with respect to the camera optical axis. The aim is to capture the shape of the pattern induced by possible surface damages (Figure 3.b), due to rolling contact fatigue effects.
The image areas framed in Figure 3 correspond to specific regions of interest (ROIs) where dedicated processing is performed. The elaboration approach chosen for ROI1 is a 2D image processing based on blob analysis [24], while that ROI2 is a 3D image processing based on triangulation [25].

3. 2D-surface analysis

Groups of connected pixels within a closed boundary define a blob: we thought reasonable to argue that the position, number and dimension of these blobs could be directly correlated to wear- or fatigue-induced small surface dips, which lead to non-uniform backscattering of the light diffused by L1.

Figure 4. Samples representative of the different surfaces studied: (a) specimen 'A', (b) specimen 'B'; and (c) specimen 'C'.

To verify this hypothesis, the specimens shown in Figure 4 have been considered. The first one (specimen 'A') is a 'post-mortem' sample, made of rail steel UIC 900 A, which rolled 800 kcycles against a specimen made of Superlos® in a lubricated rolling contact fatigue test. The surface presents a serious damage, with dips and crests due to both wear and RCF effects. The second one (specimen 'B') is a specimen made of rail steel UIC 900 A, and has rolled 800 kcycles against a specimen made of ER8; the surface is more uniform than the one in Figure 4.a, but exhibits a certain level of roughness. The third one (specimen 'C') is made of rail steel UIC 900 A, before being tested on the cyclic contact tests, i.e., in the absence of surface damage.

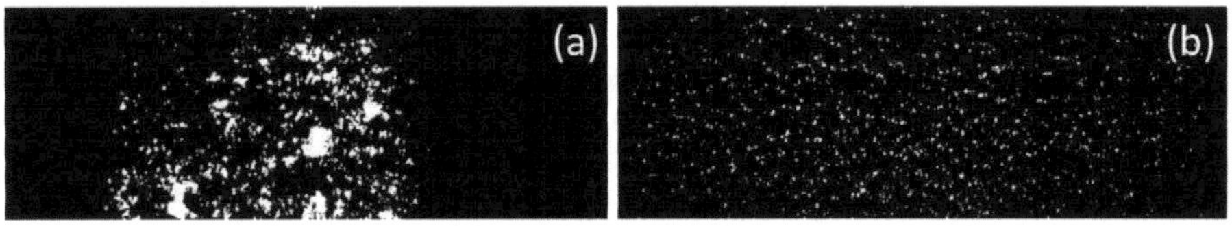

Figure 5. Example of the elaboration based on blob analysis carried out on specimens 'A' and 'B': (a) blobs detected on the ROI1 on specimen A; (b) blobs detected on the ROI1 of specimen B.

Figure 5 shows the results of the elaboration performed on the corresponding ROI1s of samples 'A' and 'B'. Blobs have been obtained after suitable thresholding, to binarize images. The processing has been performed in the LabVIEW™ environment, using the NI IMAQ Vision suite of functions. The threshold has been set to 100 (on 8-bit black/white images).

In specimen 'A' (Figure 5.a), the blobs are concentrated in a confined region of the ROI, and their position reflects the topology of the surface damage, which is localized in the central part of the sample, where the contact between specimens takes place. In specimen 'B' (Figure 5.b), the blobs are uniformly located over the whole ROI, according to the fact that, in this sample, the wear process has increased the surface roughness, without localized material spalling. In addition, the two images differ in both the number and the area of the blobs. The blob analysis performed on the surface of the specimen C (not shown) reveals very few small blobs in accordance to the fact that the surface has not yet been subjected to any wear.

Based on these observations, the following parameters have been considered in the analysis: the blob density, the dispersion of the blob centres of gravity and the average blob area. For each analysed frame, the blob density is defined as follows:

$$\mathbf{d_B = N_B/A_1} \tag{1}$$

where N_B is the number of the blobs detected in ROI1 and A_1 is the area of ROI1 (in pixels).

The dispersion of the blob centres of gravity is defined as follows:

$$\mathbf{s_{xf} = ST.DEV}\{B^i\}_c \tag{2}$$

where $\{B^i\}_c$ represents the row coordinate of the center of gravity for each ith blob belonging to column c, and ST.DEV$\{B^i\}_c$ is the standard deviation of the $\{B^i\}_c$ values with respect to the image columns.

The average blob area is calculated using the following formula:

$$a_B = \frac{1}{N_B}\sum_{i=1}^{N_B} a_i \tag{3}$$

where a_i is the area of each blob in the image. The values calculated for d_B, s_{xf} and a_B for specimens 'A', 'B' and 'C' are shown in Table 1. Here, the maximum values of both d_B and s_{xf} are observed in specimen 'B' according to the fact that N_B is higher than for specimen 'A' and 'C' and that the blobs are present over the whole ROI, which is confirmed by Figure 5.b. The value of a_B in specimen 'A' reflects the fact that the blobs area span a wide range, resulting into high values of both the average value and the standard deviation.

Besides the above parameters, the following index has been also defined:

$$r_B = \frac{\sum_{i=1}^{N_B} a_i}{A_1} \tag{4}$$

Parameter r_B represents the ratio between the image area occupied by the blobs and the total area of ROI1 and is an estimation of the damage ratio. For specimens 'A', 'B' and 'C' the values of r_B are 6.9 %, 2.6 % and 0.0024 %, respectively.

Seven specimens have been tested for 2D surface analysis: all them are made of the same rail steel UIC 900 A, and have been tested for the same number of cycles (800 k). They only differ for the type of rolling contact fatigue test (dry or wet/lubricated), and, when the type of test is the same, for the type of counter-rolling wheel steel in the test. For each specimen, a 2 s video has been recorded, and each frame has been elaborated.

Following completion of the processing of the temporal frame sequence, the average values of the above parameters over all frames, i.e. D_B, S_X, A_B and R_B are calculated.

Table 1. The values of d_B, s_{xf} and a_B in specimens 'A', 'B' and 'C'.

	d_B [px^{-1}]	s_{xf} [px]	a_B [px]
specimen 'A'	2.9×10^{-3}	100.7	24.2 ± 80.4
specimen 'B'	4.9×10^{-3}	165.0	5.3 ± 4.2
specimen 'C'	1.2×10^{-5}	5.7	2

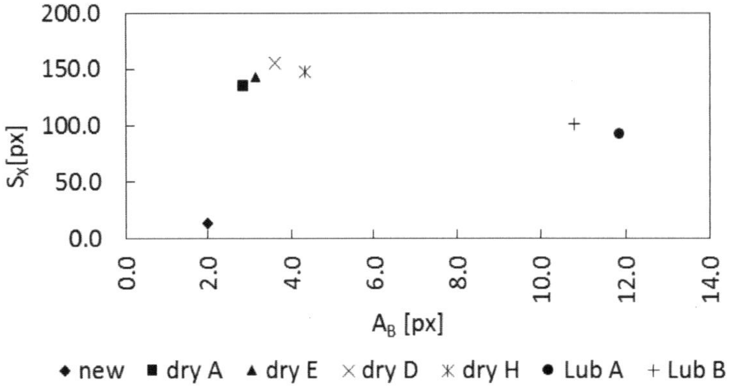

Figure 6. Blob dispersion S_X *vs* the single blob average area A_B.

Figure 6 shows a plot of the blob dispersion S_X *vs* the average blob area A_B for all the tested specimens. Each point in the graph represents a different specimen: 'new' is specimen 'C' in Figure 4.c; 'Lub A' underwent a wet test and rolled against a wheel made of ER8; 'Lub B' is the specimen 'A' in Figure 4.a; 'dry A' and 'dry H' underwent dry tests and rolled against wheels made of ER8; 'dry D' and 'dry E', underwent dry tests and rolled against wheels made of Superlos®. From Figure 6 it is evident that 'new' exhibits small values of both A_B and S_x. All the 'dry' specimens exhibit small average blob areas (between 3 and 5 px), but with markedly higher blob dispersion (about 150 px). The LubA and LubB specimens are both characterized by values of A_B between 10 and 12 px, and S_X values of about 100 px.

The observation of the plot in Figure 6 highlights that new specimens, 'dry' and lubricated ones, are all well separated from each other. This result is in accordance with the literature which shows that wear effects are dominant in dry tests, while fatigue effects are dominant in lubricated tests [19, 22, 23].

Figure 7 shows a plot of the density of blobs D_B *vs* the average blob area A_B for all the tested specimens. Here, specimens that underwent a dry test are clearly separated from the ones that underwent a lubricated test, principally thanks to very different values of A_B, while, for what concerns D_B, the two types of specimens overlap. In particular, for 'dry' specimens the density of blobs assumes different values, from 500 px^{-1} to 2000 px^{-1}, while the 'lubricated' ones show only high values of D_B, included between 1500 px^{-1} and 2000 px^{-1}. As expected, the new specimen is characterized by a value of D_B very close to zero, and is therefore distinguishable from 'dry' and 'lubricated' ones.

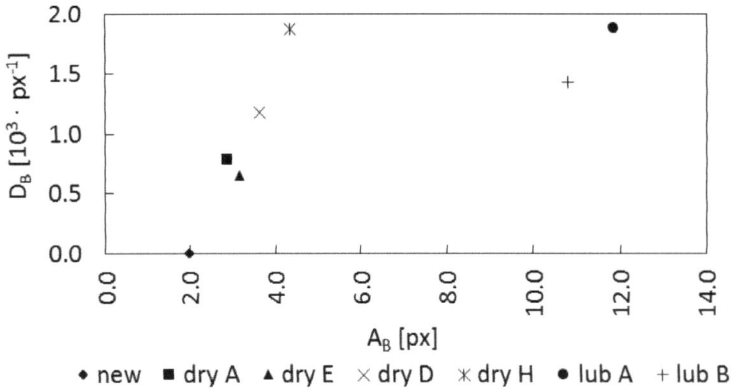

Figure 7. Density of blobs D_B *vs* the single blob average area A_B.

By Figure 6 and Figure 7, it appears that A_B and S_x are suitable indexes to distinguish between new specimens and different types of surface damage, while DB is a more variable quantity and does not unambiguously identify a single type of specimen.

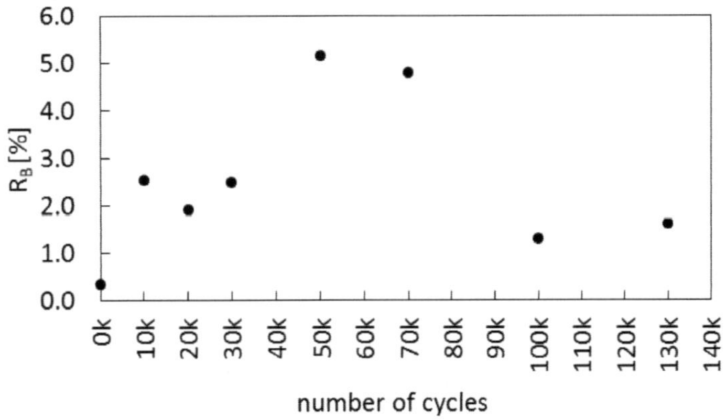

Figure 8. Time evolution of the damage ratio R_B.

The system has also been characterized in relation to its ability to monitor the temporal evolution of specific specimens during dry rolling contact fatigue tests. In particular, a specimen made of ER8 has been monitored by acquiring videos after pre-defined numbers of cycles. Figure 8 shows the time evolution of the area ratio R_B *vs* the number of cycles. R_B increases rapidly after 10 kcycles, remains fairly constant up to 30 kcycles. It increases further after 50 kcycles, until 70 kcycles. Finally, after 100 kcycles, R_B decreases. This result is in accordance with the research presented in [23], in which RCF tests have been performed to study how RCF and wear proceed and interact. These results confirm that wear is a concurrent phenomenon, which removes material layers from the surface and consequently affect ratchetting evolution. In particular, wear has a role in stabilizing the strain field for a high number of cycles, and prevents long crack formation. This is also supported by Figure 8, which shows how the damage ratio R_B for a higher number of cycles is lower than for an intermediate number of cycles.

4. 3D surface analysis

4.1. The 3D measurement performed on ROI2

The aim of this procedure is to retrieve the depth of a deformation of the surface in correspondence of the points illuminated by laser L in Figure 9, which correspond to L2 in Figure 1. This measurement has been thought to be of interest, especially in combination with the 2D description of the specimen resulting from the blob analysis. Optical triangulation has been used to perform this task. The principle is well known, and can be easily understood by looking at the geometry shown in Figure 9. Here, points L and C are the exit and the entrance pupils of the laser and the camera optics respectively. Considering a surface point A, its conjugate at the camera plane is A'. The variation along the depth coordinate Z is represented by distance h: the laser ray $\overline{LA}$ impinges the surface at point B, which is imaged at point B' at the image plane. The relationship between the depth variation h and the ray shift $y' = \overline{A'B'}$ is obtained considering that triangles $\widehat{BAD}$ and $\widehat{CAL}$ are similar, hence the following relationship holds:

$$\frac{y}{h} = \frac{d}{s} \tag{5}$$

where $y = \overline{DB}$ and $d = \overline{CL}$.

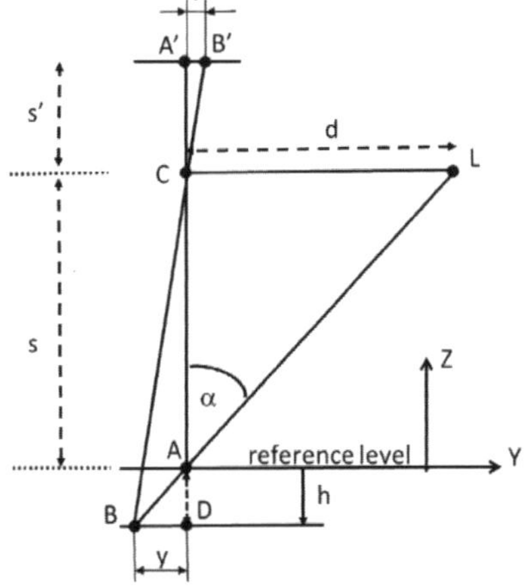

Figure 9. Optical geometry of the 3D measurement approach.

In addition, the relationship between y and y' is as follows:

$$\frac{y'}{y} = \frac{s'}{s-h} \quad \textbf{(6)}$$

where h can be a negative number.
Combining Eq. (5) and Eq. (6), and solving for h, we obtain the following formula:

$$h = \frac{s^2 \cdot y'}{d \cdot s' + y' \cdot s} \quad \textbf{(8)}$$

Eq. (8) states that, by measuring the deformation y' along Y of the line profile, it is possible to estimate the surface depth along the Z coordinate (provided that parameters s, s' and d are known). Values y' can be measured by evaluating, for each image column of index c, the centre of gravity of the corresponding grey levels $\{l_r\}$ by means of the following relation:

$$b_c = \frac{\sum_r l_r \cdot r}{\sum_r l_r} \quad \textbf{(9)}$$

As an example, Figure 10 shows the result of this procedure when applied to the grey level values shown in Figure 3.b: each point represents a centre of gravity of coordinates (c, b_c). By multiplying each value b_c by the pixel size, the corresponding y' value is obtained (in mm). The estimate of parameters s, s' and d has been performed by means of a suitable calibration procedure, that is presented in the following Section 4.2.

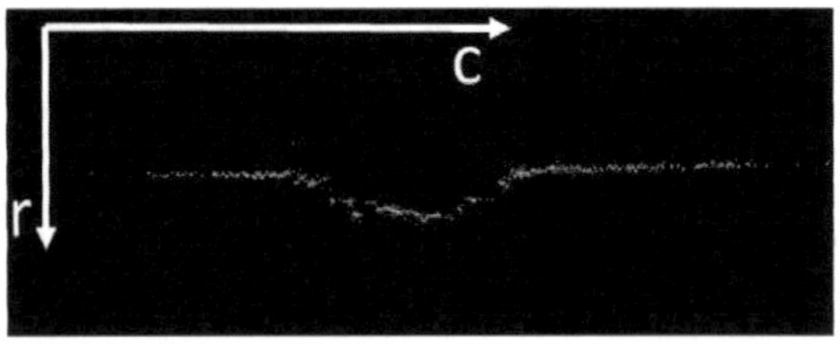

Figure 10. Row-shift at each column of the image in Figure 3.b

4.2. Procedure for estimating the 3D measurement parameters
The estimation of parameters s, s' and d has been accomplished as follows. Solving Eq. (8) for parameter y' we obtain:

IOP Conf. Series: Journal of Physics: Conf. Series **778** (2017) 012007 doi:10.1088/1742-6596/778/1/012007

$$y' = \frac{h \cdot s' \cdot d}{s(s-h)} \tag{10}$$

In Eq. (10), s and s' are linked by the well-known Gaussian formula:

$$\frac{1}{s} + \frac{1}{s'} = \frac{1}{f} \tag{11}$$

or:

$$s' = \frac{S \cdot f}{S - f} \tag{12}$$

where f is the focal length. Looking at Figure 9, d can be derived as:

$$d = s \cdot tg(\alpha) \tag{13}$$

Substituting (11) and (12) in (10), the following relation is obtained:

$$y' = \frac{h \cdot f \cdot s \cdot tg(\alpha)}{(s-h) \cdot (s-f)} \tag{14}$$

Eq. (14) has been used to estimate parameter s. To this aim, a plane calibration master has been oriented parallel to plane XY, and has been moved at known positions $\{h_n\}$ along –Z, using a sliding table with a resolution of 0.01 mm, over a range of 2.7 mm. The value h_0 has been set to correspond to the position of point A in Figure 9. At each position n, the laser pattern projected by L2 has been acquired. Using Eq. (9), the centres of gravity $\{(c, b_c)\}_n$ have been estimated with sub-pixel accuracy; the corresponding values $\{y'\}_n$ have been computed, and their average $E[\{y'\}_n]$ over the image columns has been taken as a value representative of the profile position at distance h_n.

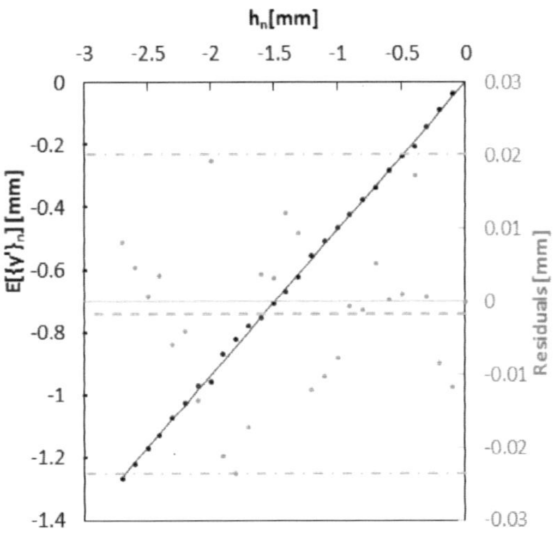

Figure 11. Calibration curve of the 3D laser slit.

Using Eq. (14) for different $(E[\{y'\}_n]-h_n)$ pairs, and at known values of focal length f and of angle α, led to the estimate of parameter s equal to 226.30 mm, using a least square method with an overall standard uncertainty of the model equal to 0.011 mm. From Eq. (12) we obtained $s' = 64.18$ mm and using Eq. (10) we evaluated $d = 376.63$ mm.
The curve shown in Figure 11 plots the relationship between $E[\{y'\}_n]$ and h_n. Black symbols (left vertical axis) represent the set of $(E[\{y'\}_n]-h_n)$ couples; their interpolation (continuous line) is well approximated by a straight line, which is expected when considering that, in Eq. (10), $h \ll s$. Symbols in grey (right vertical axis), represent the residuals, i.e. the differences between experimental points

$E[\{y'\}_n]$ and interpolated ones at each h_n value. Their distribution along Z turns to be compatible with zero, with a correlation coefficient of 0.0045, demonstrating that the residuals are stochastically distributed.

4.3. System performance

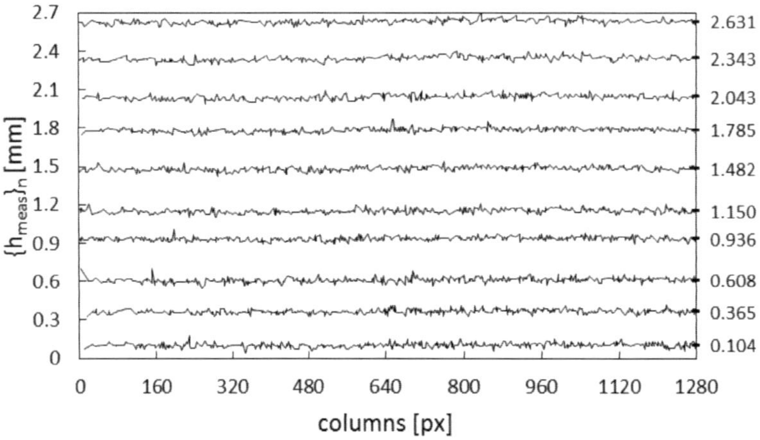

Figure 12. Values of h measured along the image columns for known positions h_n.

A number of experiments have been carried out to test the 3D measurement. First, the values obtained for parameters s, s' and d by means of the calibration described in Section 4.2 have been used in Eq. (8), to estimate the quality of the measurement. To this aim, a master plane has been positioned along the measurement interval at steps of 0.1 mm, and new images have been acquired and elaborated by using the triangulation formula of Eq. (8). As an example, Figure 12 presents the plots of measured values $\{h_{meas}\}_n$ along the image columns at each known position h_n for a starting value of n equal to 0 mm up to 2.7 mm at steps of 0.3 mm. The average value $E[\{h_{meas}\}_n]$ calculated for each $\{h_{meas}\}_n$ signal is also reported (right axis).

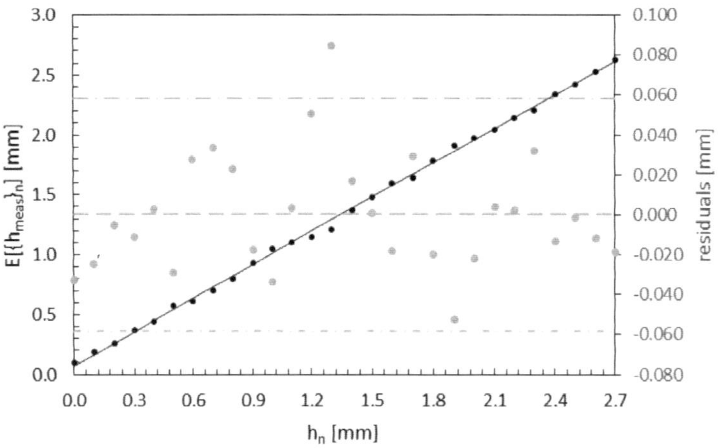

Figure 13. Plot of the average $E[\{h_{meas}\}_n]$ of measured values $\{h_{meas}\}_n$ vs known positions h_n.

Values $E[\{h_{meas}\}_n]$ are plotted *vs* values h_n in Figure 13 (black dots). The linear interpolation results into an angular coefficient of (0.9542 ± 0.0075), with a linear correlation coefficient of 0.999. Along the right vertical axis, in grey, the residuals are presented. Their average value is compatible with zero and the corresponding linear coefficient is equal to 0.33, which demonstrate a stochastic distribution of the residuals.
An example of the measurement performed on specimen 'B' of Figure 4.b is reported in Figure 14.a: the stripe projected by the laser source L2 is visible on the surface of the specimen. Figure 14.b shows the image acquired while the specimen was rotating at fs=500 rpm; the resulting measurements are shown in Figure 14.c.

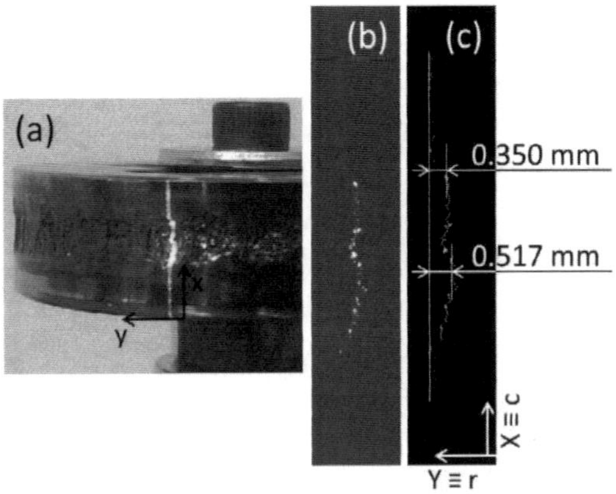

Figure 14. Example of a 3D measurement, performed on images acquired at a speed of 377 fps. (a) the line profile projected on the specimen surface; (b) the corresponding image of the deformed line profile; (c) measured values $\{h_{meas}\}_n$ along the image columns (X coordinate).

5. Conclusions

We have demonstrated the feasibility of a novel set-up for the on-line monitoring of rolling contact fatigue tests on wheel/rail specimens, based on a high-speed camera and on two laser sources, able to perform micro- and 3D macro-profile measurements, as well as angular position measurements. A prototype has been realized and its performances studied. The system is now installed in the test bench for rolling contact fatigue tests and is yields new information about the different types of damages, such as wear and fatigue, and the behaviour of different materials during the whole lifetime of the specimens.

6. References

[1] D. F. Cannon, K. O. Edel, S. L. Grassie, and K. Sawley 2003 *Fatigue Fract. Eng.M.* **26** 865.
[2] U. Zerbst, and S. Beretta 2011 *Eng. Fail. Anal.* **18** 534.
[3] Z. Liu, W. Li, F. Xue, J. Xiafang, B. Bu, and Z. Yi 2015 *IEEE Trans. Magn.* **51** 6201907
[4] Z. Li, M. Molodova, A. Núñez, and R. Dollevoet 2015 *IEEE Trans. Ind. Electron.* **62** 4385
[5] Z. Liu, F. Li, B. Huang, and G. Zhang 2014 *J. Opt. Soc. Am. A* **31** 1721
[6] Q. Li, and S. Ren 2012 *IEEE Trans. Instrum. Meas.* **61** 2189
[7] L. Jie, L. Siwei, L. Qingyong, Z. Hanqing, and R. Shengwei 2009 *IEEE International Symposium on Industrial Electronics (Seoul)* p 769
[8] U. Olofsson, and T. Telliskivi 2003 *Wear* **254** 80
[9] G. Zhou, C. He, G. Wen, and Q. Liu 2015 *Tribol. Int.* **91** 160
[10] C. Kammerhofer, A. Hohenwarter, and R. Pippan 2014 *Wear* **316** 101

[11] D. I. Fletcher, and S. Lewis 2013 *Wear* **298-299** 57
[12] J. Zhang, M. Korsten, and P. Regtien 2003 *Proc. XVII IMEKO World Congress (Dubrovnik)* p 1960
[13] G. A. Al-Kindi, and B. Shirinzadeh 2007 *Int. J. Mach. Tool. Manu.* **47** 697
[14] E. Kayahan, H. Oktem, F. Hacizade, H. Nasibov, and O. Gunogdu 2010 *Tribol. Int.* **43** 307
[15] Y-K. Fuh, K. C. Hsu, and J. R. Fan 2012 *Opt. Laser Eng.* **50** 312
[16] Z. Gao, and X. Zhao 2012 *Opt. Laser Eng.* **50** 668
[17] J. Sukuraman, S. Soleimani, P. De Baets, V. Rodriguez, K. Douterloigne, W. Philips, and M. Ando 2012 *Wear* **296** 702
[18] G. Sansoni, P. Bellandi, and F. Docchio 2011 *Meas. Sci. Technol.* **22** 075302
[19] A. Mazzù, C. Petrogalli, and M. Faccoli 2015 *Wear* **322-323** 181
[20] M. Lancini, I. Bodini, D. Vetturi, S. Pasinetti, A. Mazzù, L. Solazzi, C. Petrogalli, and M. Faccoli 2015 *ACTA IMEKO* **4** 66
[21] A. Mazzù, L. Solazzi, M. Lancini, C. Petrogalli, A. Ghidini, and M. Faccoli 2015 *Wear* **342-343** 22
[22] G. Donzella, M. Faccoli, A. Ghidini, A. Mazzù, and R. Roberti 2005 *Eng. Fract. Mec.* **72** 287
[23] G. Donzella, M. Faccoli, A. Mazzù, C. Petrogalli, and R. Roberti 2011 *Wear* **271** 408
[24] T. B. Moeslund 2012 BLOB Analysis *Introduction to video and image processing – Building real systems and applications* ed. Undergraduate topics in computer science (London: Springer Verlag) pp 103 – 116
[25] G. Sansoni, M. Trebeschi, F. Docchio 2009 *Sensors* **9** 568

Acknowledgments
We acknowledge the invaluable assistance of Prof. A. Mazzù, Dr. M. Faccoli and Dr. C. Petrogalli in the preparation of the specimens, and in the design of the experiments. We are also very grateful to Lucchini RS – Italy, that provided material and technical support, and wish to thank Mr. S. Bonometti, Mr. B. Tratta and Mr. G. Coffetti for their support in the experimental activities.

IOP Conf. Series: Journal of Physics: Conf. Series **778** (2017) 012008 doi:10.1088/1742-6596/778/1/012008

Laser doppler and radar interferometer for contactless measurements on unaccessible tie-rods on monumental buildings: Santa Maria della Consolazione Temple in Todi

M Gioffré[1], N Cavalagli[1], C Pepi[1], M Trequattrini[2]

[1]Department of Civil and Environmental Engeneering, University of Perugia, via G. Duranti, 93-06125, Perugia, Italy

[2]BQT S.n.c, Via Curtatone, 26-06073, Solomeo di Corciano, Perugia, Italy

E-mail: massimiliano.gioffre@unipg.it

Abstract. Non-contact measurements can be effectively used in civil engineering to assess the variation of structural performance with time. In the last decades this approach has received considerable interests from researchers working in the field of structural health monitoring (SHM). Indeed, non-contact measurements are very attractive because it is possible to perform non intrusive and non destructive investigations even being at a significant distance from the targets. Within this context, contactless measurements of the tie-rod vibrations in the Santa Maria della Consolazione Temple in Todi (Italy) are presented in this paper. In particular, laser vibrometer and radar interferometer measurements are used to estimate natural frequencies and mode shapes. This information is crucial to obtain the tensile axial force in the tie-rods, which can be used as an indicator of structural integrity or possible failure. Furthermore, a novel approach is proposed where drones (Unmanned Aerial Vehicles) can be successfully used to improve the effectiveness and the accuracy of the experimental activities.

1. Introduction

Structural Health Monitoring (SHM) of historical constructions is becoming one of the most important tools for safeguarding and preserving monuments and cultural heritage [1–4]. Italy is one of the countries characterized by the presence of a considerable number of historical constructions, which require important maintenance programs, significant restoration activities, and planning of emergency strategies in case of natural hazards.

The state of a structure can be altered in time by natural degradation of materials, severe isolated events such as earthquakes and thunderstorms, and background environmental loads such as synoptic winds or traffic. Within this context, SHM can provide crucial information about the ability of a structure to accomplish a required function in time. Modern static and dynamic monitoring techniques can be used both to provide information about the main mechanical and geometrical parameters used in the structural identification and to detect damage occurrence in time. In the last years, one of the most appealing monitoring technique is the one based on structural vibration [5–9], aimed to assess natural frequencies, mode shapes and associated damping ratio.

Figure 1. Temple of Santa Maria della Consolazione in Todi.

Tie rods are simple structural elements that have been commonly used throughout the centuries to improve stability of old masonry buildings. These elements, different in material, size and shape, are used as an effective connection in the structural system constituting elements, especially lateral walls or springing of arches and vaults. Stability of the whole construction and/or local structural elements is directly connected to the magnitude of the tensile axial force carried out by the tie-rods. A change in time in this tensile force might be associated to some kind of mechanical failure causing a new internal forces distribution among the structural elements. As an example, cracks in the masonry walls or differential settlements, which can compromise the stability and the integrity of the whole building, can initiate this process of change in the internal forces in the structural elements, including the tie-rods.

The evaluation of the axial internal force time histories in the tie-rods, e.g. with periodic measurements, can be used as an effective monitoring tool in order to predict structural failures. This evaluation has to take into account all the uncertainties in the geometrical and mechanical properties due to both the use of hand made iron elements and the different constraint boundary conditions.

Measurements of tie-rods vibrations can be obtained using contact and non-contact vibration techniques. The development of non-contact measurement systems has received a considerable interests from researchers in the last decades (e.g. [10, 11]). Non-contact measurements are very attractive because it is possible to perform non intrusive and non destructive investigations even being at a significant distance from the target.

In this paper, contactless measurements of tie-rod vibrations in the Santa Maria della Consolazione Temple in Todi (Italy) are presented. Laser Doppler Vibrometer and Radar Interferometer tests are compared, showing both ambient vibration and induced free vibrations. A novel usage of Unmanned Aerial Vehicles (UAV), commonly known as drones, to support the experimental tests is also described together with a set of thermografic survey in order to estimate the tie-rod temperature during the measurements. This information is crucial for estimating the tie-rod axial tensile force variation with time.

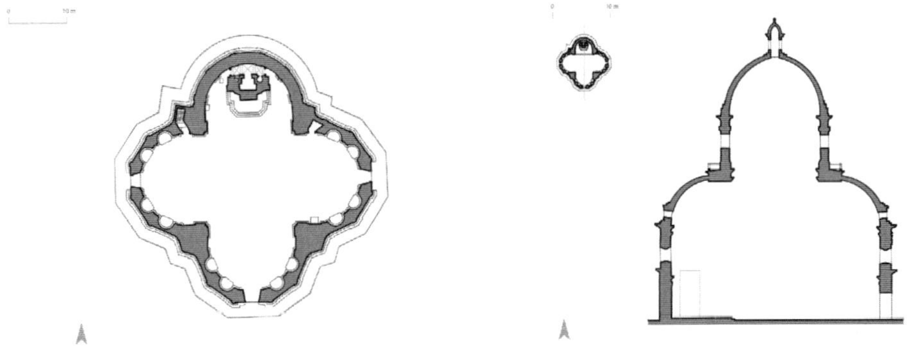

Figure 2. Temple of Santa Maria della Consolazione in Todi: plan and longitudinal section (Courtesy of Biondini & Corradi Associati - Studio di Ingegneria e Architettura.)

2. Santa Maria della Consolazione Temple

2.1. The Temple

The Santa Maria della Consolazione Temple (Figure 1), one of the most important monumental buildings in Umbria Region, in Central Italy, is located in the South West Area of Todi's city. It is a Renaissance style church and it was built between the 16th and the 17th centuries. Although direct evidence is lacking, the first design of the structure can be assigned to Bramante [12].

The church consists of a greek cross plan, with 42 m and 48 m dimensions, with a tall central dome, which rises to a total height of about 50 m from the temple floor, and four semi-shells on the lateral sides. The dome consists of a single shell with ogival shape [13], it rests on a circular drum, is surmounted by a lantern and reinforced with ribbs. Figure 2 shows the plan and the longitudinal section of the temple carried out after a recent architectural survey.

The main bearing structure consists of 4 pillars with an height of about 25 m and a thickness of about 2 m. The base of the pillars lays on a square central area of 25 m^2. The North apse, the first to be built, is semi-spherical, while the others are polygonal. The apse walls thickness is of about 2 m. The foundations reach a depth of 4 m. All the structure is made of ashlar and mortar. The exterior of the temple is made of the white limestone blocks from the local quarries. The dome and the apses shells are covered with lead sheets.

Since its construction, dated back to 1509, the Temple revealed structural problems and local collapses. The first consolidation work began at the end of 1800s. Four tie-rods were installed: each element was connected to the spring line of the vaults and directly fixed to stonework by bolts and plates. At the same time, a retaining wall was built in the South area. In the second half of the 20th century drilling holes in concrete foundations and cement grouting were carried out in order to strengthen foundations and masonry walls. A drainage wall was also completed in order get the water out from the structure [14]. In the eighties it was discovered that the cause of instability was a landslide on the South West side, due to the differential sliding of clay and sands. Because of this reason, a consolidation work began in 1990, where 232 micropiles and 257 large diameter piles were built [15]. After the recent Umbria-Marche earthquake in 1997, the temple was subjected to other phases of strengthening. Seismic safety was improved by confinement measures on the apses and dome internal side and on the drum external side.

2.2. The tie-rods

The four tie-rods in the Santa Maria della Consolazione Temple (Figure 3(a)) appeared for the first time in a picture dated back in 1890 (Figure 3(b)). During the years these metallic elements

(a) (b)

Figure 3. (a) Bottom view of the central dome. (b) Historical picture (1890) of the tie-rods.

experienced failures and replacements. The North and the East tie-rods were replaced in 1905 and 1959, respectively; the former after breaking, while the latter because of the instability problems involving the whole structure.

As of today, the geometrical and mechanical properties of the tie-rods are unknown so that every analysis has to consider uncertainties on several parameters such as bar length, cross section size, mass density, elastic properties, and support conditions. Furthermore, each of the tie-rods is equipped with a tightening device made with iron eyelets in order to adjust the rod tension in time to accommodate slight changes in the structural geometry. The position and the masses of the tightening systems are further unknown parameters to be considered in future structural global and element identification.

3. Experimental Tests

A laser vibrometer and a radar interferometer have been used to measure tie-rod vibration time histories. This choice finds its main reason on the difficulty in placing commonly used contact sensors, such as accelerometers and strain gauges, on the unaccessible tie-rods located at the height of about 16 m. The obtained measurements can be used to estimate natural frequencies and mode shapes that are directly related to the tensile axial force.

3.1. Brief description of the measuring system

The laser doppler vibrometer used in this work is a Polytec OFV 3001 - OFV 303, controller and sensor head(Figure 4(a)), with resolution 0.25 nm, constant in the whole frequency range from 0 Hz to 250 kHz.

In order to improve signal quality it was decided to use reflecting targets to be applied on the target positions in the unaccessible tie-rods. This issue was overcome designing a suitable drone (UAV) equipped with a carbon fiber mechanical arm as shown in Figure 5(a). Figure 5(b) shows the reflecting target, correctly positioned by the drone, and the vibrometer measurement point on the South tie-rod.

The main features of the IBIS-FS ground based radar interferometer are summarized in Table 1.

3.2. Experimental layout

Figure 6 (a) shows the laser vibrometer (blue circles) and the radar interferometer (white squares) positions used in the experimental measurements. It is worth noting that the target surfaces are located at a height over 16 m (Figure 6 (b)). A drone (UAV) was used to place

(a)

(b)

Figure 4. (a) Laser Doppler vibrometer system. (b) Radar Interferometer.

(a)

(b)

Figure 5. (a) Drone at work. (b) Reflecting target on the South tie Rod.

Table 1. Features of IBIS-FS 010-14-000301 ground based radar interferometer.

Center frequency f_0	17.2 GHz
Frequency Bandwidth	200 MHz
Range resolution ΔR	0.75 m
Maximum EIRP Power P_0	26 dBm
Maximum unambiguous displacement	4.36 mm

reflecting targets on the unaccessible tie-rods avoiding time consuming and costly scaffolding or moving platforms.

4. Experimental Results

4.1. Temperature measurements

The tie-rod temperature was measured using thermal images. This information is crucial to distinguish tensile force variation that can be due both to geometrical and to temperature changes. Figures 7 (a) and (b) show the thermal images used to estimate the temperature in the East and West tie-rods. Maximum, minimum and mean temperature values are summarized in

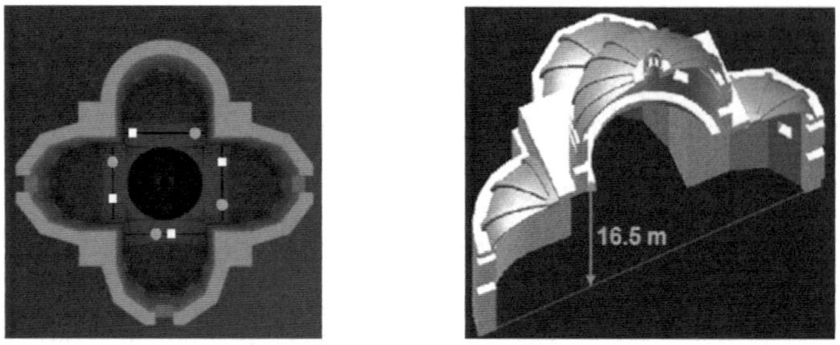

(a) (b)

Figure 6. (a) Plan view of the laser vibrometer (blue circles) and radar interferometer (white squares) positions. (b) Tridimensional view of the North Apse with the tie-rod height.

Table 2.

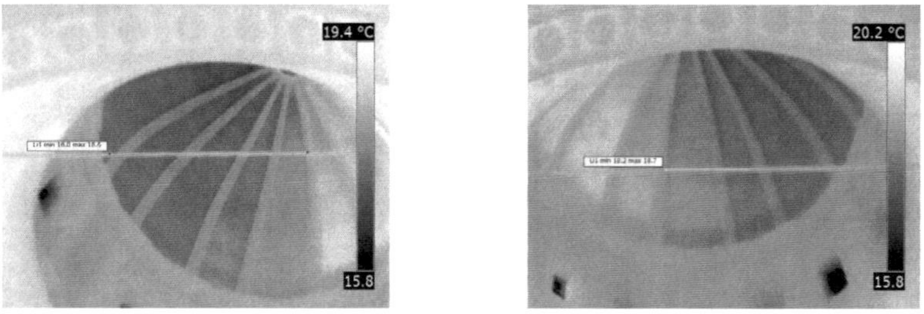

(a) (b)

Figure 7. Thermal image of the East (a) and West (b) tie-rod.

Table 2. Maximum, minimum and mean temperature values in the four tie-rods.

Tie Rod	Tmin[°C]	Tmax[°C]	Tmean[°C]
North	18.40	19.00	18.70
East	18.00	18.60	18.30
South	18.60	19.00	18.80
West	18.20	18.70	18.45

4.2. Ambient vibration measurements
Ambient vibrations were recorded in all the tie-rods with a sampling rate set to 200 Hz. Figure 8 shows the displacement time histories obtained by the laser vibrometer and radar interferometer in the East tie-rod.

Figure 9 shows the power spectral densities estimated from 100 s of the displacement time histories in the East (upper panels) and North (lower panels) tie-rods. The left and right panels report estimates from the laser vibrometer and radar interferometer measurements, respectively.

4.3. Free Vibrations

Free vibrations were induced giving a force pulse with a thin rope. Then, 100 s displacement time histories were recorded with sampling rate set to 200 Hz. A drone (Figure 5(a)) was used to overpass the thin rope over the four tie-rods. The thin rope was suitably linked to the drone using a special connection system in order to avoid instability problems.

Figure 10 shows the displacement time histories obtained by the laser vibrometer and radar interferometer in the West and North tie-rods, respectively. Left panels refer to laser vibrometer, while right panels report radar interferometer measurements. The different shape of the North

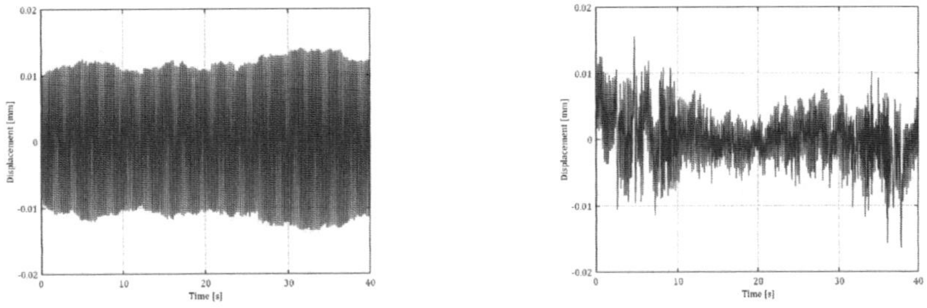

(a) (b)

Figure 8. East tie-rod displacement time histories: (a) laser vibrometer (b) radar interferometer.

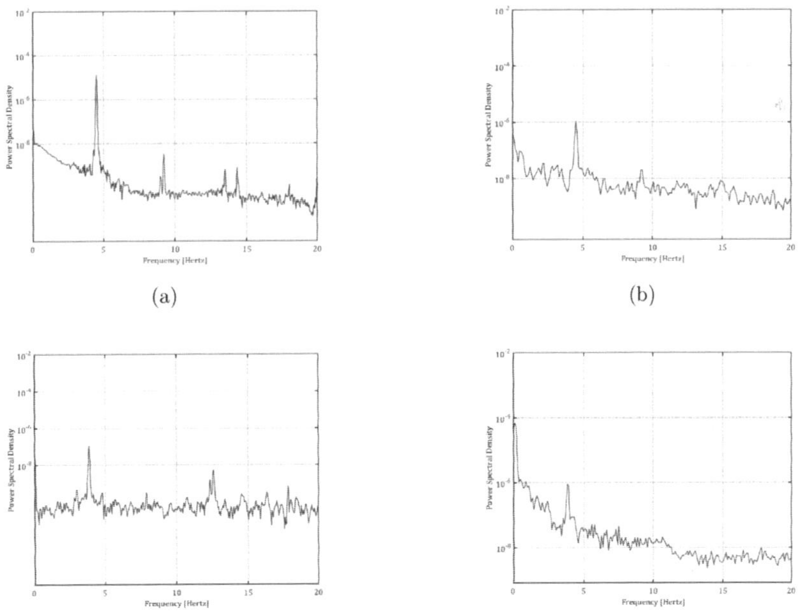

(a) (b)

(c) (d)

Figure 9. Power spectral densities estimated from the laser vibrometer (left panels) and radar interferometer (right panels) measurements in the East (upper panels) and North (lower panels).

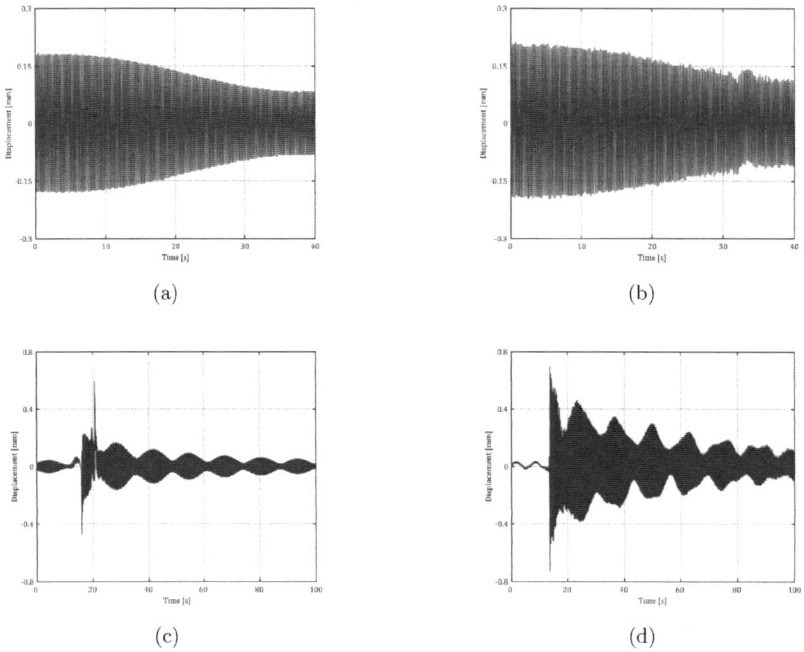

Figure 10. Displacement time histories from the laser vibrometer (left panels) and radar interferometer (right panels) in the East (upper panels) and North (lower panels).

tie-rod vibrations is due to three hanging candle holders playing the rule of pendulous.

Figure 11 shows the power spectral densities estimated from 100 s of the displacement time histories in each tie-rod. The left and right panels report estimates from the laser vibrometer and radar interferometer measurements, respectively.

More information can be obtained by the radar interferometer since it is possible to record simultaneous displacements in a finite number of sample points located along its line of sight. As an example Figure 12 (a) shows the measured point locations on the North tie-rod, while Figure 12 (b) reports the displacement time histories at these locations.

Figure 13 (a) and 13 (b) show the maximum and minimum displacement (blue) obtained at the different measured positions, together with the tie-rod geometry at two time istants (red).

4.4. Tie-rods natural frequencies

The tie-rods natural frequencies can be easily estimated by performing a peak picking in the frequency domain (e.g. Figures 9 and 11). The first natural frequency value for each tie-rod is summarized in Table 3 both for laser vibrometer and radar interferometer data. The results comparison demonstrate that both the instruments provide the same values in the East and West tie-rods. A slight difference, less than 2%, is found for the North and South elements.

5. Conclusions

Laser vibrometer and radar interferometer vibration measurements of unaccessible tie-rods on the Santa Maria della Consolazione Temple in Todi were presented and compared in this paper. It was shown that both measuring techniques give satisfactory results in terms of the first natural frequencies and can be an effective alternative to contact sensors such as accelerometers and

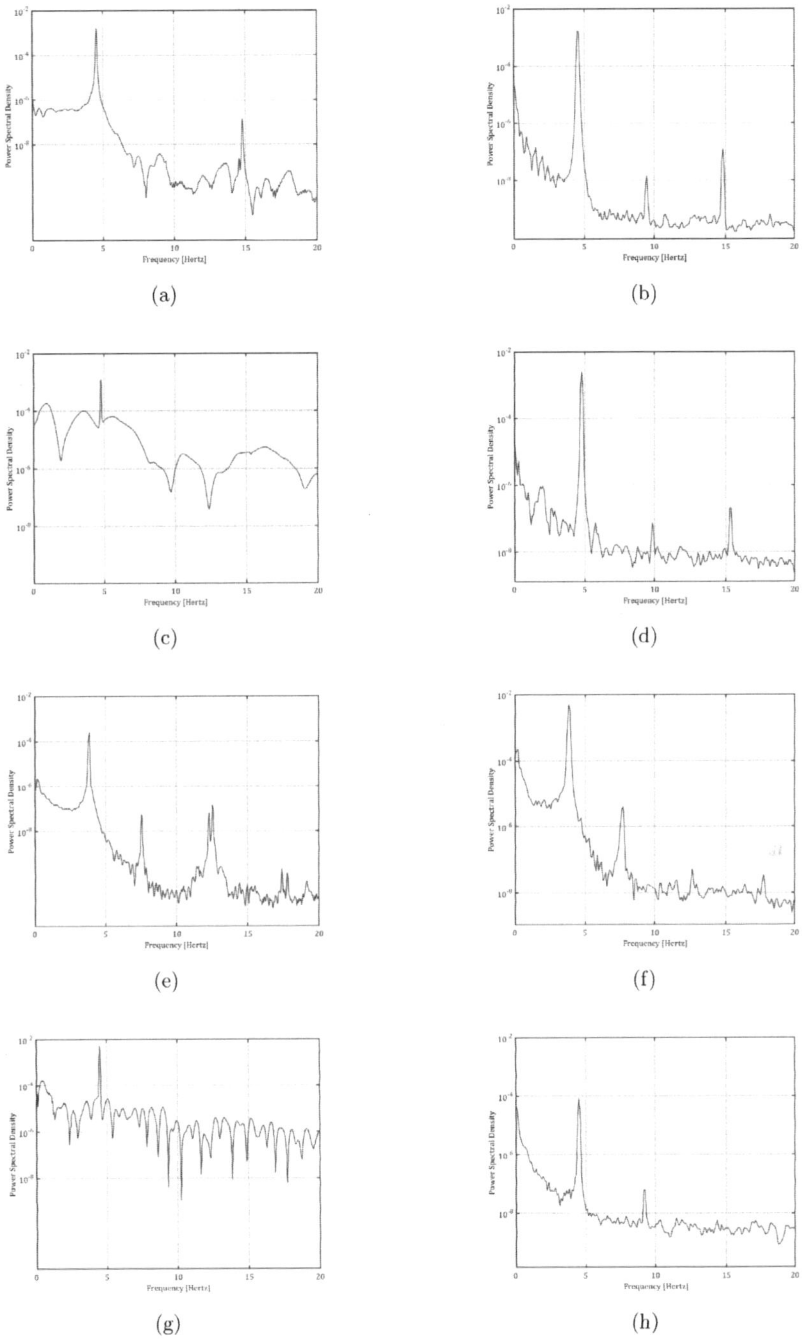

Figure 11. Power spectral densities estimated from the laser vibrometer (left panels) and radar interferometer (right panels) measurements in the South, West, North, East tie-rods (top to bottom panels).

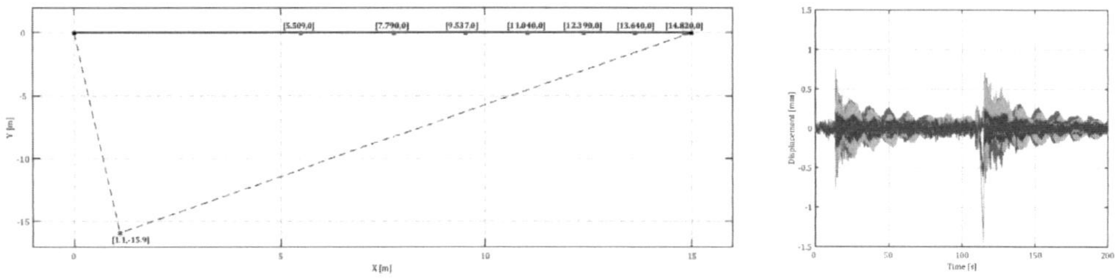

(a) (b)

Figure 12. (a) Sampling locations and radar interferometer position used for the North tie-rod. (b) Displacement time histories recorded by the radar interferometer at the sampling points on the North tie-rod.

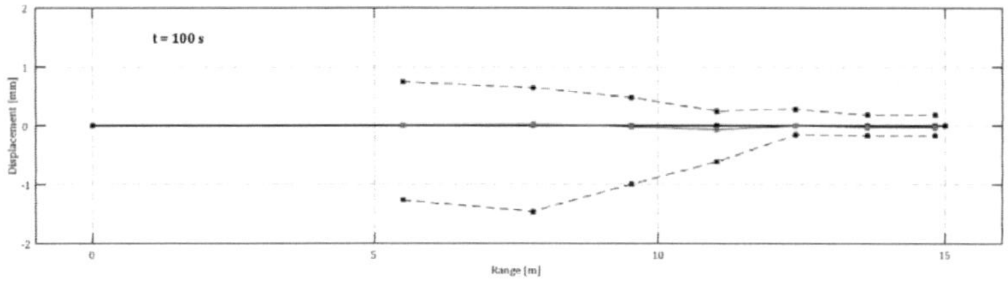

(a)

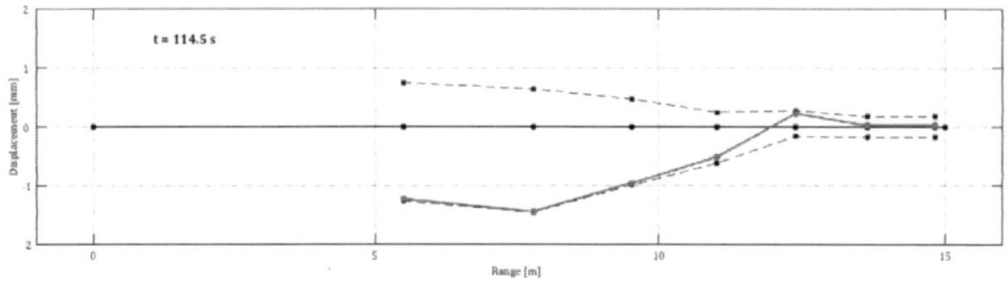

(b)

Figure 13. Maximum and minimum displacements at the sampled points (blue) of the North tie-rod recorded by radar interferometer. The red line shows the tie-rod shape at two time instants: (a) 100 s and (b) 114.5 s.

Table 3. Tie-rod natural frequencies obtained by laser vibrometer and radar interferometer.

Tie Rod	Vibrometer (Hz)	Interferometer (Hz)
East	4.785	4.785
West	4.875	4.875
North	4.541	4.491
South	3.875	3.809

strain gauges. Furthermore, a novel methodology was proposed to improve the effectiveness and the accuracy of the experimental activities by the use of drones (Unmanned Aerial Veichles) to place reflecting targets on suitable locations and to position thin ropes for inducing free vibrations of the structural elements.

6. Acknowledgments

These Authors gratefully acknowledge "ETAB La Consolazione" for the logistic support and the interest demonstrated to this research activity. These Authors are also in debt to Carlo Intotaro, Droinwork APR Aereal Service, for his continuous support with designing, manufacturing and flying the drones. Finally, these Autors wish to thank Ing. Francesco Bianchi at the CIRIAF (University of Perugia) for the help with the experimental tests.

References

[1] Bartoli G, Chiarugi A and Gusella V 1996 Monitoring systems on historic buildings: the Brunelleschi Dome *Journal of Structural Engineering ASCE* **122**(6) 663-673

[2] Gioffré M, Gusella V and Cluni F 2008 Performance evaluation of monumental bridges: testing and monitoring "Ponte delle Torri" in Spoleto *Structure and Infrastructure Engineering: Maintenance, Management, Lyfe-Cycle Design and Performance* **4**(2) 95-106

[3] Fanelli M 2015 Importance of Monitoring, Behavior Analysis, and Diagnostic Techniques for Preservation of Great Historical Structures *International Journal of Architectural Heritage* **1**(9) 2-6

[4] Ottoni F and Blasi C 2015 Results of a 60-Year Monitoring System for Santa Maria del Fiore Dome in Florence *International Journal of Architectural Heritage* **1**(9) 7-24

[5] Ramos LF, De Roeck G, Lourenço PB and Campos-Costa A 2010 Damage identification on arched masonry structures using ambient and random impact vibrations *Engineering Structures* **32**(1) 146-162

[6] Saisi A, Gentile C and Guidobaldi M 2015 Post-earthquake continuous dynamic monitoring of the Gabbia Tower in Mantua, Italy *Construction and Buildings Materials* **81** 101-112

[7] Cavalagli N, Gioffré M and Gusella V Structural monitoring of monumental buildings: The Basilica of Santa Maria Degli Angeli in Assisi (ITALY) *Proceedings of COMPDYN 2015 - 5th ECCOMAS Thematic Conference on Computational Methods in Structural Dynamics and Earthquake Engineering* 2410-2422

[8] Gentile C, Guidobaldi M and Saisi A 2016 One-year dynamic monitoring of a historic tower: damage detection under changing environment *Meccanica* doi: 10.1007/s11012-016-0482-3 Article in press

[9] Ubertini F, Comanducci G and Cavalagli N 2016 Vibration-based structural health monitoring of a historic bell-tower using output-only measurements and multivariate statistical analysis *Structural Health Monitoring* **15**(4) 438-457

[10] Gentile C (2010) Deflection measurement on vibrating stay cables by non-contact microwave interferometer *NDT & E International* **43**(3) 231-240

[11] Rossi G, Marsili R, Gusella V and Gioffré M (2002) Comparison between accelerometer and laser vibrometer to measure traffic excited vibrations on bridges *Shock and Vibration* **9**(1-2) 11-18

[12] Nofrini U 1970 *Il Tempio di Bramante a Todi* Res Tudertinæ 11, Comune di Todi, Todi, Italy.

[13] Cavalagli N, Gusella V (2015) Structural Investigation of 18th-Century Ogival Masonry Domes: From Carlo Fontana to Bernardo Vittone *International Journal of Architectural Heritage* **9**(3) 265-276.

[14] Battistoni L 2012 *Il Tempio di Santa Maria della Consolazione in Todi. 1508-1670*, 2nd Edition, ETAB - Associazione Pro Todi, Todi, Italy.

[15] Mariani M 2006 *Trattato sul consolidamento e restauro degli edifici in muratura*, DEI Tipografia del Genio Civile, Rome, Italy.